The Battle of Priorities

Politics versus The Planet

by
Paula B. Johnson

Dear Esteemed Reader,

Thank you immensely for choosing this book to join your collection. We imagine that you've already embarked on an exploration of ideas within these pages, and we couldn't be happier about it!

Now, if you find yourself chuckling, pondering, or even debating with the words in front of you, we'd absolutely love to hear about it. If you can spare a few moments to pen down your thoughts in a review, we would be as delighted as a dictionary on a spelling bee!

An Amazon review would be excellent - but hey, we're far from picky. Whether it's a scribble on the back of a grocery list, a tweet, or even a message in a bottle (though that might take a while to reach us), your feedback is gold.

Writing a review might not be as fun as a spontaneous dance-off, but we promise it'll bring grins to our faces, warmth to our hearts, and incredibly valuable insights to future readers.

With Gratitude,

Bo Bennett, PhD Publisher

Archieboy Holdings, LLC.

Table of Contents

Introduction 1

Chapter 1: Understanding the Global Climate Crisis 3

Chapter 2: The Politics of Environmental Science........................... 12

Chapter 3: The Economics of Climate Policy 23

Chapter 4: Lobbyists and Climate Policy.. 34

Chapter 5: Public Opinion and Climate Policy................................ 45

Chapter 6: Case Study: The Paris Agreement 55

Chapter 7: Navigating the Political Landscape for
Climate Action 65

Chapter 8: Green Parties and Climate Politics 76

Chapter 9: The Future of Climate Politics....................................... 87

Conclusion 98

Introduction

Climate change is indisputably one of the most urgent crises humanity is currently facing, with scientific consensus identifying the pressing need for substantial interventions to mitigate its devastating impacts. However, the complexities of the relationship between politics and climate change often present significant impediments to decisive action. This book endeavors to elucidate this complex relationship and to illustrate how politics can often hinder, rather than advance, progress in addressing climate change.

The policy decisions and political debates revolving around climate change will shape the trajectory of our planet's future. This book will delve into an exploration of how politics and policy intersect with environmental science, economics, public perception, and global cooperation, each of which has its own distinct role in driving or hindering climate action. You will be introduced to the layers of influence that politics wields on environmental science, understand the interplay between climate policy economics and international trade, and discover the role of lobby groups and green parties in climate politics.

The objective is not merely to inform, but to engender an understanding of how these seemingly disparate entities are interlinked, and how they consequentially shape our world's efforts to combat climate change. We hope that by comprehending these intricacies, you will gain a clearer grasp of the political obstacles that hinder climate action, and more importantly, how we can overcome them. This understanding is imperative for anyone - whether a policy maker, an activist, or simply a concerned citizen - as we navigate toward a sustainable future.

Chapter 1:
Understanding the Global
Climate Crisis

A s we delve into the heart of this book, it's essential to first understand the root of our exploration: the worldwide climate predicament. The Global Climate Crisis, as it's universally recognized, is a direct consequence of escalated greenhouse gas emissions, primarily carbon dioxide, methane, and nitrous oxide, which capture and maintain heat in the Earth's atmosphere, inciting a global temperature rise (NASA, 2021). This warming effect, largely propelled by human-induced activities such as fossil fuel consumption, deforestation, and industrial processes, leads to both conspicuous and subtle shifts in global climate patterns.

Consequences extend from rising sea levels to increased incidences of extreme weather events and threatened biodiversity. Inextricably tied to these physical changes are a chain of socio-economic impacts such as food and water insecurity, public health threats, displacement of populations, and destabilization of economies. While this paints a somewhat grim reality, "Understanding the Global Climate Crisis" is not just about recognizing the causes and consequences. It's also a prerequisite to explore the interlinkages of climate considerations and political decisions, which form the core narrative of this book.

Facts and Figures

The understanding of the depth of the global climate crisis isn't complete without the facts and figures that demonstrate the extent of the issue.

Global temperatures are soaring, polar ice caps are melting, wildfire frequencies are rising, and extreme weather events are becoming more common. These realities underline the urgency of the crisis we're dealing with.

According to NASA, the average global temperature on Earth has risen about 0.8° Celsius (1.4° Fahrenheit) since 1880 (NASA, 2020). Even more alarming, two-thirds of this warming has happened since 1975—a rise of 0.15-0.20°C (0.3-0.4°F) per decade. Global warming does not occur evenly across the world. The Arctic region, in a phenomenon known as Arctic amplification, is heating up twice as fast as the global average (Serreze & Barry, 2011).

The rise in temperature is leading to the melting of ice caps and glaciers worldwide. The rate of Antarctica ice mass loss has tripled in the last decade (IMBIE, 2018). This consequence of global warming, along with other factors, contributes to the ongoing rise in global sea levels.

Research indicates that sea levels worldwide have been rising at a rate of 3.3 millimeters (.13 inches) per year in the last couple of decades (Church & White, 2011).

Another worrying trend directly linked with the global climate crisis is the increase in the frequency and intensity of wildfires. Studies show that the number of large fires in the western U.S. and Alaska has increased since the early 1980s (Westerling et al., 2006). Long fire seasons and a rise in average wildfire sizes can be attributed to human escalation of greenhouse gases.

In parallel with the increase in temperatures and wildfires, we're experiencing more extreme weather events. According to a report by the Intergovernmental Panel on Climate Change (IPCC), it's observed that heavy precipitation events have increased in some parts of the world and further increases are expected in the high and some of the middle latitudes (Stocker et al., 2013). This has led to increases in flooding and associated damages.

The global climate crisis is also leading to more intense heatwaves. Research suggests that we're experiencing more and longer heatwaves compared to 50 years ago (Meehl et al., 2004). These devastating trends have dire consequences, from heat-related deaths and illnesses to amplified drought conditions, placing additional stress on water resources and agriculture.

Climate change also has a pronounced influence on global ocean cycles. Evidence indicates that human-induced warming is disrupting usual patterns of El Niño and La Niña events, causing more intense cycles and changing global rainfall and temperature patterns (Cai et al., 2014). This presents a variety of potential threats to global ecosystems and human societies.

An additional factual concern relating to the climate crisis is related to carbon dioxide (CO_2) emissions. The concentration of CO_2 in our atmosphere in May 2021 was the highest monthly average ever recorded, at 419.13 parts per million (NOAA, 2021). The ramifications of these increased levels of CO_2 can be seen in higher global temperatures, melting glaciers, and rising sea levels.

It's not just the rising temperatures and melting glaciers that are of concern, the indirect effects of climate change are just as worrying. This includes the surge in vector-borne diseases like malaria and dengue fever, as mosquitoes expand into newly warm areas (Ryan et al., 2019).

Similarly, indigenous communities whose livelihoods depend on the natural environment are being unduly affected by the changing climate (Ford et al., 2014).

Another indirect effect is the rise in climate refugees. According to the Internal Displacement Monitoring Centre (IDMC), in 2019 alone, about 23.9 million people were displaced in 140 countries due to severe weather events (IDMC, 2020). These climate migrants often face harsh living conditions and human rights challenges.

Finally, the economic costs of the climate crisis can't be discounted. The Global Commission on Adaptation estimates that the economic damages of climate change could reach $1 trillion annually (Global Commission on Adaptation, 2019). These costs will only increase if significant strides are not made in reducing greenhouse gas emissions and adapting to the changes already taking place.

In the face of all these facts and figures, the science is clear. Climate change is no longer a remote problem, it's here and happening now. Understanding the statistics behind climate change is the first step towards taking action in mitigating its effects on our planet.

Causes and Consequences: The causes and consequences of climate change are numerous and interconnected. A rise in the emission of greenhouse gases, primarily carbon dioxide and methane, is the fundamental cause of the current global climate crisis. These gases are primarily produced by human activity, including the burning of fossil fuels for energy, deforestation which eliminates vital carbon sinks, and agricultural activities which release methane (IPCC, 2014).

Greenhouse gases trap heat in the Earth's atmosphere causing global temperatures to rise, resulting in a process known as global warming. The consequences of global warming are far-reaching, culminating in a wide variety of alterations in our climate system. We observe this through recurring episodes of much more severe heatwaves, forest fires, droughts, and storms.

These physical manifestations of climate change also lead to serious human and ecological consequences. It influences the availability of resources such as water and food, exacerbating existing socio-economic inequalities (Union of Concerned Scientists, 2021). It also threatens biodiversity as habitats are altered and species are forced to migrate or face extinction.

Sea-level rise, another consequence of global warming due to the thermal expansion of the ocean and melting polar ice caps, is on track to displace millions of people within this century, creating a new class of climate refugees. Coastal urban locations will also bear the brunt of expenses from damages caused by increased flooding and the necessity of adapting their infrastructure (Climate Central, 2019).

Climate-induced health issues are another grave concern. The World Health Organization (WHO) estimates that the direct health impacts of extreme weather events and the indirect effects of changing patterns of infection, nutrition, and air quality represent a large proportion of the increased mortality and illness due to climate change (WHO, 2018).

These causes and consequences of climate change underscore the urgency of climate action and the need for a coordinated global political response.

The Role of Politics in Climate Change

As we delve deeper into the global climate crisis, it's fundamental to understand the significant role politics plays in climate change matters. Politics, in this sense, refers to the process by which leaders and governments make decisions, set policies, allocate resources, and achieve public goals. Politics influences every aspect of climate change, including awareness, advocacy, policy-making, regulation, enforcement, and funding for climate-related initiatives.

From a global perspective, the politics of climate change is a question of negotiation between the nations of the world. Every nation has a role to play in mitigating climate change, and the policies they devise have implications for the entire globe. The United Nations Framework Convention on Climate Change (UNFCCC) governs the international political response to climate change. This body establishes legally binding commitments for developed countries to reduce greenhouse gas emissions (Bodansky, 2010).

However, there is a constant fight over responsibility and the division of labor between developed and developing countries. Developed countries have historically contributed more to global greenhouse gas emissions, while developing countries will arguably suffer more from the impacts of climate change. Balancing these complex demands represents a significant political challenge.

The United States, as one of the world's largest greenhouse gas emitters and a leading global power, plays a particularly crucial role in climate politics. Unfortunately, the country's approach to climate change has often been erratic and inconsistent, largely due to political polarization.

Within the U.S, climate change has become a highly partisan issue, with attitudes toward it now serving as a political identity marker (McCright & Dunlap, 2011). This political divide has resulted in inconsistent climate policy, with the U.S. rapidly changing course on climate initiatives with each election cycle.

A striking illustration is the contrast between the Obama administration, which championed the Paris Agreement and a transition toward renewable energy, and the subsequent Trump administration, which withdrew from the Paris Agreement and rolled back many environmental regulations in the name of economic growth.

Politics influences not only the adoption of climate change policy but also its implementation. The decision to enforce or ignore environmental regulations often comes down to political will. Without this will, policy initiatives aimed at managing and mitigating climate change may fall short of their potential.

Further, public funding for climate research and mitigation projects often depends on political priorities. As administrations and legislative majorities shift, so too does federal funding for climate-related projects. Such fluctuations can disrupt ongoing research and complicate long-term planning efforts.

Moreover, lobbying by powerful industries can significantly influence political responses to climate change. Fossil fuel and other high-emitting industries can wield immense political influence to protect their interests, often to the detriment of sound environmental policy (Oreskes & Conway, 2010).

Despite the many challenges posed by politics, the realm of policy and politics also holds the greatest potential for sweeping, systemic change. Comprehensive and sustained action on climate change needs systemic and structural transformations, and these can only happen through political mechanisms.

Democracy itself can be a powerful tool in this fight. Citizens can put pressure on their governments to act on climate change, can elect leaders who prioritize the issue, and can hold these leaders accountable for their promises. Therefore, public opinion and civic participation constitute a crucial dimension of climate politics.

In conclusion, without understanding the role of politics, we cannot fully grasp the complexities of climate change nor map our path towards effective solutions. The political realm presents both significant obstacles and promising opportunities for combating climate change. Navigating these dynamics will be a critical part of our journey toward a more sustainable future.

Global Politics and Climate are deeply interconnected facets of our world. Arguably, how nations collectively approach the climate crisis is one of the most significant political issues of our time. Without a unified approach, initiatives to mitigate climate change can become inefficient and ineffective (Milkoreit, Hulme & Moore, 2018).

Nationally, countries might pledge their commitment to reducing carbon emissions and transitioning to renewable energy. Yet, when these commitments are examined on a global stage, they may be found lacking in ambition or even in direct contradiction to other political actions. For instance, nations could pledge to Paris Agreement goals while simultaneously expanding fossil fuel industries. This duplicity presents a significant political challenge. It illustrates not only the complexity of climate politics but also the limitations inherent in the management of a global crisis within a system of nation-states (Falkner, Stephan & Vogler, 2019).

International politics also frequently revolve around power dynamics, with certain countries wielding more influence over climate negotiations than others (Kahiluoto et al., 2019). This imbalance can lead to inequitable outcomes, disproportionately affecting those most vulnerable to the impacts of climate change. If we consider, for instance, developing nations struggling for economic growth and meeting their energy needs, the mandates from developed nations to restrict their carbon emissions can be seen as unfair. The politics around climate aid and technology transfer further compound these disparities (Keohane & Victor, 2016).

It can also be noted that partisan politics and ideology often shape climate change policy. Policies dealing with climate change can differ greatly based on the party in power, particularly in nations with significant fossil fuel interests, such as the U.S., Russia, or Saudi Arabia (Lockwood, 2020). Moreover, conflicts of interest, political corruption, and short- term electoral cycles can slow down the implementation of long-term climate change solutions, contributing to the global delay in climate action.

To conclude, global politics and climate are closely intertwined, and understanding this relationship is crucial to addressing the climate crisis effectively. The world needs a transformative approach in global politics that can rise above national self-interest and partisan divides to make way for broad-based, ambitious, and equitable climate action. Only then can we hope to avert the worst impacts of climate change and ensure a sustainable future for all (Fröhlich & Knieling, 2020).

The United States Politics and Climate have developed a volatile relationship over the past decades. Politics has played an instrumental role in shaping national climate change response in the

United States (Leiserowitz, Maibach, Rosenthal, Kotcher, Bergquist, Ballew, Goldberg, & Gustafson, 2020). There exist significant disparities among political parties in terms of acknowledging and addressing the threat that climate change poses.

The Republican Party's views have veered towards climate skepticism or denial, while the Democratic Party has been more inclined to acknowledge and address climate change (Dunlap, McCright, & Brouwens, 2016). These partisan views concoct a deeply polarized American society, often hindering the necessary advancement of robust climate change policies.

Government administration also influences the United States' stance on climate change significantly. Notably, the Trump administration distanced the nation from international climate agreements such as the Paris Agreement, frustrating global efforts towards climate mitigation (Arora, 2019). Conversely, the Obama administration had actively endorsed such global cooperations.

The influence of interest groups has also played a significant role in shaping the U.S. political landscape on climate change. Powerful industries, particularly fossil fuel industries, have utilized their economic influence to lobby against climate change mitigation policies, fearing potential detrimental effects on their respective sectors (Brulle, 2018).

This engenders a complex interplay between politics, economic interests, and climate change. In essence, the politics surrounding climate change in the United States is driven by a wide array of factors ranging from partisan ideologies, administrative policies, to the influence of powerful industries. It is clear that the political landscape in the U.S. significantly influences both the national and global response to the climate crisis.

Chapter 2:
The Politics of Environmental Science

The intersection of politics and environmental science is a complex web of truth, bias, and conflicting interests. Environmental science, despite its essential role in informing climate policy, is easily manipulated or discounted due to political pressure and bias (Braun, 2014). This political interference can manifest in many ways, from gathering data with inherent biases to interpreting this data in ways that suit particular agendas.

Tragically, this politicization often breeds denial of the substantial and irrefutable evidence validating climate change. Consequently, science, a discipline dedicated to the impartial pursuit of knowledge, becomes tainted with political interests and agendas. Climate denial, backed by manipulated data interpretations, are potent political tools, chiefly used to undermine or delay climate action, thus preserving status quo at the expense of our environment (Oreskes & Conway, 2010). The stunting impact of such denial on necessary policies and measures is amply clear; we see an alarming trend of stagnated progress and promising reforms getting hopelessly mired in controversy and debate, thereby jamming the gears of change.

The Use and Misuse of Environmental Science

Environmental science stands as the guiding force that offers insight into our world's ongoing climate crisis. Among its tools are carefully-honed methods of data collection, analyses, and interpretation that help us comprehend the pace and magnitude of climate change (Klein, 2014).

With unequivocal accuracy, environmental science has painted a clear picture of a world under threat, comprehensively piecing together multiple facets of a complex issue. However, how these scientific findings are used and, at times, misconstrued plays a significant role in shaping worldwide climate policies - a reality that won't surprise anyone who's been following the climate debate over the years.

Sound environmental policy decisions ought to be steeped in well-rounded, rigorous scientific understanding to ascertain the best possible paths forward for our planet. From reductions in greenhouse gas emissions to the conservation of natural habitats, science offers invaluable insights into what steps must be taken to limit the adverse effects of climate change (Carrington, 2016).

However, the misuse of environmental science becomes apparent when evidence-based pursuits of climate mitigation are conflicted by political cross-currents with a vested interest in preserving the status quo. Blatantly ignoring or downplaying scientific consensus on climate matters, these political entities foster doubt and perpetuate misinformation, leading to a collective inertia that stalls substantial progress (Aaron-Morrison, 2017).

Beyond the outright dismissal of scientific consensus, there's also the subtle yet pervasive issue of cherry-picking scientific data. This tactic manipulates the narrative to fit a specific viewpoint, proving harmful to developing comprehensive, informed climate policy. It distorts the public's perception of the problem at hand, leading to polarization and consequently, policy inaction (Carrington, 2016).

Moreover, the selective promotion of certain climate mitigation strategies over others—often driven by political or economic interests—can skew our understanding of what constitutes a truly effective response to climate change. Such bias can inadvertently suppress or de-emphasize potentially more successful strategies from emerging and taking their necessary place in the climate policy arena (Klein, 2014).

Such biased manipulation undermines not just the credibility of science, but also the critical role that science ought to play in forming sound environmental policies. The distortion of science by politics disrupts the crucial dialogue regarding climate change, creating a dysfunctional dynamic where misinformation or manipulative narratives thrive, influencing public opinion and policy decisions (Aaron-Morrison, 2017).

In the interest of safeguarding our planet and future generations, the rhetoric around climate change can't afford to be a game of political football. Rigor, transparency, and openness to revision are eminent characteristics of good science, and these values should carry over to the world of policy-making (Carrington, 2016).

Transcending partisan lines, reputable environmental science advocacy must observe an unflinching commitment to facts and figures. Stripped of biases and manipulation, science has the power to inform, educate, and shape productive conversations about our planet's future. It is a tool to be used, not misused, in our pursuit of sustainable policies (Klein, 2014).

Without a doubt, this misuse of environmental science in the political landscape casts a shadow on the potential of science-based policy-making. Yet, it equally behooves us to underscore the significant strides we've made thus far—a testament to the use of environmental science. From the adoption of renewable energy technologies to policies aimed at conserving our precious ecosystems, credible, well-executed science has undoubtedly informed successful action (Carrington, 2016).

It's essential to recognize that the fight against climate change is not merely a scientific endeavor—it's a social, economic, and political one too. Environmental science provides the foundation, but it's the responsibility of our societies and governments to build on this through informed, decisive action (Aaron-Morrison, 2017).

In conclusion, the use and misuse of environmental science within the realm of politics are a delicate, double-edged sword—one that requires vigilance, discernment, and considerable nuance to navigate. With a heightened awareness of these issues, we are better equipped to advocate for scientific integrity, pursuit of knowledge, and informed policy- choices that align with our collective vision of a sustainable future. (Klein, 2014).

As we delve further into the nexus of environment, science, and politics, we will see the ways in which the correct use of scientific data can shape our response to the global climate crisis, just as we will uncover how its misuse can stymie and stall progress.

Bias in Environmental Science is a crucial factor that more often than not, sidetracks true knowledge about the climate crisis. Scientists strive for objectivity when conducting their research. However, bias, whether conscious or unconscious, can influence the outcome of research and even pervade its dissemination to the public (Harding, 2015).

To begin with, it is necessary to differentiate between biases arising from the researcher's perspective and those that emerge from political pressure. For instance, a researcher may, subtly or unconsciously, favor specific methodologies or interpret data in line with their previous experiences or beliefs. However, such bias differs from that which is introduced when a researcher is pressured to design or interpret studies in line with a political agenda (Kearney & Levine, 2016).

Regrettably, the reality in environmental science is that funding bodies, which are often governmental or corporate entities, might give rise to biased research. Relying on financial support from sources that have a specific stance on climate change can threaten objectivity in research. Consequently, some research findings may overstate or understate environmental problems to align with the interests of the funding entity (Freudenburg, Gramling, & Davidson, 2008).

In addition to funding sources, the communication of environmental science findings to the public can be swayed by political bias. Media outlets, politicians, and those with vested interests may emphasize, de- emphasize, or even distort scientific findings to suit their narratives (Shehata & Hopmann, 2012). There is a tendency for the media to provide 'balanced' coverage, which implies giving equal weight to climate change advocates and skeptics, despite the scientific consensus leaning heavily towards the reality of climate change. This practice can mislead the public and skew the perception of what is scientifically established.

In conclusion, a rigorous and objective approach is needed in both the execution and communication of environmental science to identify and mitigate bias. As consumers of scientific information, we should be wary of potential bias, scrutinizing research sources, the funding behind them, and the way results are presented by the media. Greater transparency about potential sources of bias is needed to foster trust in scientific findings and ensure informed decision making about climate change (Wilson, 2018).

Interpreting the Data is a crucial aspect that directly influences our understanding and execution of steps towards combating climate change. This process, importantly, needs to be devoid of political biases as they can skew the interpretation and potentially mislead policy decisions. As climate change is a scientifically established fact, data interpretation within the realm of environmental science should be governed by the scientific method (Kolstad, C., Urama, K., Broome, J., Bruvoll, A., Cariño Olvera, M., Fullerton, D., ... & Jotzo, F., 2014).

The interpretation of data forms the backbone of our policies for mitigating and adapting to climate change. From defining the specifics of an emission reduction target, to assessing the impact of a carbon tax, every policy decision is rooted in how the data is understood. However, issues arise when this interpretation is colored by political ideologies - for instance, conservative-leaning voices have often interpreted climate data in a way to downplay the severity of impending climate disasters, while liberal-leaning voices may highlight the urgency for action (Pielke, R., 2007). Any bias in data interpretation can lead to a misalignment of policies with the on-ground reality of climate change and potentially obstruct effective action.

Transparency in climate data interpretation can be achieved when it is an interdisciplinary effort, involving not just environmental scientists and climatologists, but also sociologists, economists, and political scientists. This allows for a comprehensive understanding that factors in the physical, societal, and economic impacts of climate change. Moreover, it is imperative to ensure the accurate communication of interpreted data to the public and policymakers. Providing open access to datasets and methodologies also enhances scrutiny and accuracy (Jamieson, D., 2014). The crux lies in imbibing a scientific temper to guide interpretation, thus enabling us to steer clear of unproductive politicization that might dilute the gravity of the climate crisis.

Politics-Driven-Denial of Climate Change

A significant obstacle in the path of grappling with climate change is the politics-driven denial of this major global issue. Climate change denial, often fueled by political ambitions, has far-reaching implications (Leiserowitz, 2008). It inhibits the development of policies geared towards addressing climate change, disrupts international cooperation on this front, and generally fosters an environment of apathy or even active resistance against addressing this growing crisis.

Political climate change denial is not a straightforward phenomenon. It comes in many forms and operates at several levels. At its base level, it involves outright denial of the existence of anthropogenic, or human- induced, climate change. However, there are more insidious forms of denial that unconsciously permeate society. These include cognitive denial (ignoring the scientific facts); interpretive denial (downplaying the severity of the situation); and implicatory denial (avoiding the ethical, political, and economic implications of climate change) (Cohen, 2011).

It's crucial to understand that climate change denial is not a matter of personality or intelligence but is rooted in ideology (Kahan, 2012). This ideology is often intertwined with political and economic beliefs, such as the fear that climate action may threaten economic growth or the autonomy of nations. For some political actors, admitting the reality of the climate crisis and committing to substantial action could require assenting to policy interventions and regulations they otherwise oppose due to their political beliefs.

Political climate change denial is most evident in the United States, where climate change has become a deeply partisan issue. The polarization of American politics has seen one party significantly downplay the implications of climate change, whereas the other party accepts the scientific consensus. This conflation of climate science with political ideology has been detrimental to efforts at climate action in the country (Dunlap & McCright, 2011).

Sadly, the denial of climate change has trickled down into the populace, and public opinions on climate change in the United States are also intensely politically polarized. Research has shown that conservatives are more likely to deny climate change than liberals, and that political party identification is one of the most significant predictors of climate change awareness and knowledge (McCright & Dunlap, 2011).

This political climate change denial extends beyond North America. In countries like Australia and Brazil, political leaders have used climate skepticism to rally their base or to justify environmentally harmful policies. Politics-driven denial of climate change is not just about denying the science, but also about denying the need for action and rejecting international cooperation (Chayes, 2020).

Such political denial is often fostered and amplified by vested interests. Industries dependent on fossil fuels, for example, have been found to lobby heavily against climate action and to fund misleading climate information (Oreskes & Conway, 2010). However, it's important to stress that this denial is not exclusive to such interests and can be found even where economic self-interest does not seem to be the clear motivator.

Denial politics heavily impacts international efforts to address climate change. International climate agreements are hampered by countries' reluctance to commit to emissions reductions targets, often driven by domestic political pressures and a denial of the severity of the problem (Hovi, Sprinz, & Underdal, 2009). This denial obstructs the comprehensive global action necessary to adequately address climate change.

On a national level, politically motivated climate change denial can have devastating outcomes. In places where such denial is prevalent, it results in a lack of effective climate legislation, prevents the implementation of necessary mitigation and adaptation measures, and can even lead to a rollback of environmental protections. The implications of such denial are profound, leading to a delay in responding to climate threats and increasing vulnerabilities to climate impacts.

Furthermore, denial politics incite a vicious cycle. Denial of climate change leads to inaction, which in turn exacerbates the climate crisis. As the situation worsens, the idea of tackling climate change can seem more daunting, leading to higher levels of denial and resistance to action (Brulle, 2014). Breaking this cycle of denial, resistance, and inaction is incumbent upon us all.

Educating the public about the genuine reality of climate change and its human causes is a vital part of this. Extensive public awareness and understanding of the issue can help counteract the impact of denial politics, applying pressure on political representatives to recognize and act on the problem. Public consciousness can be an effective antidote to politics-driven denial (Egan & Mullin, 2017).

In conclusion, the denial of climate change, spurred and maintained by political motivations, obstructs our ability to deal effectively with the global climate crisis. Recognizing and understanding this phenomenon is the first step towards confronting and overcoming it and paving the way for comprehensive action on climate change.

How Politics Fuel Climate Denial becomes clearer in instances where politicians cater and respond to the inclinations of their core base. Let's look at the United States, where climate change is often a partisan issue. These political leanings greatly influence climate denial, which thrives among certain segments of the population that lend significant political support to certain parties (Merkley & Stecula, 2020).

In fact, according to a study from Yale University, Democrats are generally more likely to believe and express concern about climate change than Republicans (Leiserowitz, Maibach, Rosenthal, & Cutler, 2018). This discrepancy can be attributed, in part, to political campaigns and strategies that see climate denial as a unifying bond among constituents or, on the contrary, a divisive instrument against the opposition.

Moreover, climate denial is often amplified by vested industrial interests whose profitability hinges on lax environmental policies. Some politicians, in the pursuit of preserving jobs and economic growth, intentionally ignore or deny the severity of climate change effects (Hertsgaard & Pope, 2021). They use their political platforms to disseminate climate disinformation, contributing to public climate skepticism, which reinforces denial and hinders policy development.

However, it's not just about climate denial; it's also about climate delay. Politicians often strategically use inaction, incrementalism, and redirection to veer away from concrete climate change policies that could lead to fundamental changes benefiting the environment. While less explicit than outright denial, these political techniques are equally, if not more, detrimental as they give the illusion of action while maintaining the status quo (Anderson, 2021).

In conclusion, politics play a pivotal role in fueling climate denial through the manipulation of public sentiment and the propagation of misinformation. Additionally, through delay tactics and the defense of vested interests, politicians add to the environmental predicament that our planet faces. The interplay between politics and climate denial showcases the urgency for informed leadership, transparency, and an emphasis on science-centric policies in addressing the climate crisis.

Impact of Climate Denial In a world where information is readily accessible, the refusal to acknowledge indisputable scientific evidence of climate change is more than just a denial of science - it's a significant obstacle to the global fight against climate change. The intentional spread of misinformation can cause serious damage. It not only muddles the facts but also sows doubt in the minds of the general public, hindering necessary action and rendering well-intentioned efforts ineffectual (Farrell, 2019).

Political climate denial can manifest itself in various ways, from dismissing climate science to rejecting policy proposals that address climate change. Such refutations are often fuelled by vested interests in sectors like fossil fuels or contradictory political ideologies. Not surprisingly, climate policy has often been a contentious issue, with consensus difficult to achieve (Hoffman, 2011). This denial, and the resulting policy gridlock, further delays urgent action required to curb greenhouse gas emissions.

Differentiating causes from effects, it's crucial to recognize that climate denial can have real, tangible consequences. By maintaining the status quo or rolling back important environmental regulations, climate policy can be effectively neutered. This can result in higher greenhouse gas emissions, contributing to global warming and laying the groundwork for more intense natural disasters like the wildfires, hurricanes, heatwaves, and floods the world is witnessing (Leiserowitz, Maibach, Roser-Renouf, Feinberg, & Howe, 2013).

Furthermore, climate denial sanctions an atmosphere of indifference towards environmental protection. By influencing public perceptions, denial can lower the urgency felt towards climate action, thereby influencing sustained habits of excessive consumption and waste. For instance, climate denial tends to propagate consumer behavior that is incongruous with sustainability, such as excessive motor vehicle use, increased meat consumption, and resistance to renewable energy systems (Capstick, Whitmarsh, Poortinga, Pidgeon, & Upham, 2015).

Countering climate denial therefore, is not simply about presenting scientific facts, but also about changing narratives and influencing behavior. Unfortunately, the impact of climate denial reverberates beyond the political realm, permeating through societal values and day-to-day actions. There is a desperate need to bridge this gap between scientific consensus and public understanding, and doing so is pivotal to paving the way for effective climate action (Hoffman, 2011).

Chapter 3:
The Economics of Climate Policy

We have previously discussed how politics permeates the conversation around climate change. Now, let's delve into the economic implications of climate policies, which are often at the core of these political debates. The economics of climate policy is a vast field that encompasses cost-benefit analyses, affordability debates, and industry impacts. Contention often arises on two major fronts in this discussion: the short-term economic costs of implementing climate action, and the long-term economic benefits from mitigating the impact of climate change (Stern, 2008).

Many argue that climate action, such as reducing carbon emissions or investing in renewable energy, will impose substantial costs on economies, affecting industries and jobs. However, others contend that these actions can create new economic opportunities, promoting innovation and job growth in burgeoning sectors such as renewable energy and sustainable technologies. We must also weigh the potential disruptions to international trade and the contrasting needs of developed and developing nations. This complex interplay of economic factors significantly influences decision-making in climate policy, between immediate tangible costs and more speculative long-term benefits.

Alleviating this tension to achieve a viable economic pathway for effective climate policy is a challenge that lies at the heart of climate politics (Nordhaus, 2013).

Paula B. Johnson

Affordability Debate in Climate Policy

There is widespread agreement among scientists and policymakers on the urgency of addressing the global climate crisis. However, the debate continues on how we can balance the priorities of climate change mitigation and economic stability, leading to an ongoing discussion on the affordability of comprehensive climate policies (Heal & Millner, 2014).

One side of this debate argues that implementing new climate policies can be economically devastating, foreseeing costly infrastructure overhauls, job losses in traditional carbon-intensive industries, increased energy prices, and potential downsides for international competitiveness (Flachsland, Pahle, & Edenhofer, 2016).

Others, however, view climate policy action as a unique opportunity to transition towards a low-carbon economy, which could foster sustainable growth, create new industries and jobs, enhance energy security, and improve public health. This perspective often highlights the cost of inaction, pointing out that climate change's physical, health, and socio- economic impacts can significantly outweigh the costs of intervention (Aldy, 2019).

The cost of mitigation measures varies depending on their extent, duration, and focal points. For example, some argue that rapidly transitioning to renewable energy sources could be expensive initially but would pay off in the longer run due to decreased dependency on fossil fuels and their volatile prices (Jacobson, Delucchi, Cameron, & Mathiesen, 2018).

Adaptation measures, such as building climate-resilient infrastructures or improving hazard response systems, also necessitate substantial investment. Nonetheless, they can generate significant returns by reducing the impacts of climate-induced catastrophes on communities and industries (Ebi, Semenza, & Rocklöv, 2016).

Addressing the affordability concerns necessitates recognizing the heterogeneity of impacts across society. The cost burden may disproportionately fall on low-income households or developing nations without adequate resources to invest in mitigation or adaptation actions. Climate policy ought to incorporate strategies for equitable cost distribution (Averchenkova & Bassi, 2016).

Fiscal policies, such as carbon pricing or phasing out fossil fuel subsidies, have been proposed as cost-effective ways to encourage emission reductions. Simultaneously, these measures may generate revenue for governments, which can be invested back into the economy or used to offset any regressive impacts on lower-income households (Goulder & Schein, 2013).

There's also the aspect of 'stranded assets'. Delaying the transition to a low-carbon economy may result in traditional energy assets, such as coal mines or gas power plants, being abruptly devalued or abandoned, leading to potentially significant economic disruptions (Caldecott, Howarth, & McSharry, 2013).

Innovation can play an instrumental role in overcoming the affordability challenge. Advanced technologies in renewable energy, energy efficiency, carbon capture and storage, and climate-smart agriculture can significantly reduce the costs of action, all while opening up new growth opportunities (Popp, Hascic, & Medhi, 2011).

Considering the long-term economic benefits, the argument that effective climate policy lacks affordability can seem short-sighted. In fact, many economic models suggest that early, consistent measures in line with the

1.5°C target of the Paris Agreement are economically beneficial when considering the full societal costs, including impacts on human health and ecosystems (Stern, 2018).

Given these differing perspectives, the debate around the affordability of climate policy is more nuanced than it may seem initially. The financial considerations are important, but they must be viewed within a larger context that considers current and future social, environmental and economic impacts.

In conclusion, the costs of climate policies are indeed a crucial aspect of the broader climate debate. It is important that these costs are measured accurately and weighed against the long-term benefits and the costs of inaction. After all, what we're really dealing with is the potential cost of our survival, which is something that can't be put into dollars and cents.

Economic Costs of Climate Action turn global focus and urgency toward understanding its real impact and implications. However, there's no straightforward answer to this. A multitude of factors come into play, making an exact cost calculation challenging. Certain areas will bear the brunt more than the others, and the costs will not be equally distributed.

Nevertheless, it's essential to acknowledge that the cost of inaction far outweighs the cost of immediate action (IPCC, 2018).

Initially, transitioning to a low-carbon economy can impose significant expenses. Industries that heavily rely on fossil fuels, such as coal-mining, oil, and gas, will likely experience economic disruptions - loss of jobs and income. A significant shift like this necessitates corresponding policy measures targeted at assisting and supporting the transition. Bloomfield and Bouzarovski (2018) concluded that it would profoundly affect low- income households who spend a larger share of their budget on energy and transportation.

Moreover, climate policies that put a price on carbon--such as carbon taxes or emission trading systems--could increase energy prices. For instance, a study by the Congressional Budget Office (CBO, 2013) showed that a carbon tax in the United States would lower GDP by up to 0.9% by 2050, predominantly due to higher prices that decrease consumers' purchasing power and reduce investment.

Despite these costs, it's imperative to remember that the modelling for economic costs doesn't often account for the benefits of avoiding severe climate change. Ignoring the financial implications of environmental degradation and extreme weather patterns, rising healthcare costs due to increased diseases and allergies, and the risk of regional conflicts spurred by resource scarcity can lead to a skewed perception of actual costs (Stern, 2006).

As such, it's crucial to keep the discourse centered on striking a balance between economic considerations and ecological debt. Delaying action will only escalate the economic costs in the long run, further exacerbating social, ecological, and economic challenges. After all, economics, as a social science, shouldn't govern our decisions but aid society—not impede it—in achieving its broader goals, such as safeguarding the planet for future generations.

Economic Opportunities in Climate Action delves into the prosperous aspects of environmental preservation. Often, political debates about climate action focus on the costs, causing some sectors to neglect the vast array of economic opportunities. No doubt, pursuing climate action requires significant financial resources. However, these investments can result in considerable economic returns, offering a potential solution to the otherwise divisive debates.

One notable area of economic opportunity resides in the renewable energy industry, such as wind, solar, and hydro power. Besides being pivotal for reducing greenhouse gas emissions, these alternatives can stimulate substantial economic growth. According to the International Renewable Energy Agency, the global economy could gain USD 98 trillion by 2050 through a more rapid transition to renewable energy, representing an increase of 2.5% GDP (IRENA, 2020).

Moreover, this sector has the capacity to create numerous job opportunities. In 2018 alone, it provided 11 million jobs globally, and the figure is projected to increase (IRENA, 2019). Significant job creation can not only boost economies, but it also serves as an excellent tool for political buy-in. Rural communities and previous fossil-fuel-centric economies can be revitalized through diversification and the infusion of green jobs, transforming the traditional narrative of jobs versus environment to jobs and environment.

Climate action also presents opportunities in green infrastructure development, such as energy-efficient buildings and public transport systems, bolstering the construction industry while reducing environmental impact. For instance, the American Council for an Energy- Efficient Economy (ACEEE) estimates that energy efficiency measures could add nearly 1.3 million jobs to the U.S. economy by 2050 (ACEEE, 2020). Similarly, in sectors such as agriculture, sales of organic food and drink in the U.S. reached $50 billion in 2018, underscoring the potential for profit and employment from sustainable farming practices (Organic Trade Association, 2019).

Lastly, the innovation and commercialization of green technologies, products, and services can unlock further economic growth. Climate- smart and resource-efficient solutions are not only crucial for mitigating environmental impact, but they also represent an immense market potential, with a projection to reach $26 trillion by 2030 (Global Commission on Economy and Climate, 2018). It reveals environmental action can align with fiscal prosperity, providing a compelling argument for climate policy across political lines.

Factors Influencing Climate Policy Economics

One of the significant facets of climate policy lies within its economics. Multiple complex factors can shift the economic landscape of climate policy. In this context, we are contemplating such factors that significantly mold the economics of climate policy. Drawing from a multitude of academic sources, we'll comprehend these dynamics and the impact they have on policy-making to address climate change.

The economics of climate policy are intrinsically linked with international trade. Trade practices and agreements often carry implications for the climate, based on what goods are traded and how they're produced. During international negotiations, nations often have differing perspectives on the role trade should play in climate policy (Frankel, 2009).

While some argue that enhancing trade can help disseminate green technologies, critics contend that unchecked international trading practices can lead to excessive resource use and carbon emissions, thereby undermining efforts to reduce global warming. For instance, the United States' withdrawal from the Paris Agreement has been partly traced back to concerns about international trade impacts of climate actions (Hafner-Burton, Victor, & Lupu, 2012).

Another factor that significantly determines the economics of climate policy is the status of nations as developing or developed. By and large, developed nations have been historically responsible for the majority of greenhouse gas emissions due to industrialization (Shindell et al., 2018). Therefore, the financial burden of transitioning to a low-carbon economy is often seen as the responsibility of these nations.

However, this distinction between developing and developed nations leads to several economic and ethical dilemmas. While developed countries may possess the resources and technology to combat climate change, developing nations often struggle with the high costs of transitioning to renewable sources (Hallegatte et al., 2016).

The approach to environment protection and greenhouse gas emission control also varies markedly between developed and developing countries. While developed countries focus more on mitigation, developing countries prioritize adaptation to climate change due to immediate vulnerabilities, leading to significant variation in policy costs (Shukla et al., 2018).

Different technology adoption rates between developed and developing nations can also introduce varying costs. Advanced economies have been shifting towards clean technologies, often at a faster pace than developing nations (Fankhauser et al., 2013). This results in disparate climate responses which can complicate global climate cooperation.

The financial mechanisms supporting climate policy making also present a significant economic factor. These include grants, soft loans, and carbon credit trading systems. However, the economic viability of these mechanisms can vary significantly, influencing the effectiveness and nature of environmental policies (Stavins, 2011).

Multilateral Climate Funds (MCFs), for instance, have emerged as key financing mechanisms. These funds are designed for lower-income countries to adapt to the impacts of climate change and transition to a sustainable low-carbon development pathway. However, their effectiveness and efficiency in achieving their objectives have been questioned (Nakhooda et al., 2013).

The economics of carbon pricing is another crucial factor. Policymakers use this tool to internalize the external costs of carbon emissions. By putting a price on carbon, the system generates a powerful incentive for businesses and households to reduce emissions. However, the significant regional variation in carbon prices can impact the competitiveness of certain regions or sectors and has been a contentious subject in international negotiations (Aldy et al., 2016).

Lastly, the influence of special interest groups and the political economy also plays a crucial role in the economic aspects of climate policy.

Lobbying efforts can lead to skewed resources and subsidies which make climate policies more expensive or less effective. Mitigating these influences is a significant challenge in evolving cost-effective and fair climate policies (Oreskes & Conway, 2010).

To understand the economics of climate policy, we need to grasp these lattices of influences and see them in an interconnected way. The friction between international trade interests, developing versus developed nations, financing mechanisms, carbon pricing, and the influence of special interests cannot be overlooked. Policymakers need to address these complexities to effectuate a fair and viable economic philosophy for combating climate change globally.

International Trade Impacts play a substantial role in shaping the economics of climate policy. International commerce gives nations powerful leverage to enforce climate mitigation measures. For instance, countries may impose border adjustments in the form of tariffs to penalize others with weak environmental standards. However, such measures can incite political tension, causing nations to retaliate equally or resist compliance (Peters, 2016)

The utilization of border carbon adjustments is a controversial topic. Critics argue it can lead to a breakdown in global trade relations and create barriers to entry for developing countries that significantly rely on industrial emissions for their economic growth (Cosbey et al., 2019).

However, proponents contend that without these measures, industries in countries with stringent climate policies are disadvantaged. There is also potential for carbon leakage where companies relocate their high polluting production to countries with lax regulations, ultimately counteracting climate mitigation efforts (Cosbey et al., 2019).

Moreover, an essential question lingers; to what extent can international trade law accommodate climate change objectives? Current trade regulations under the World Trade Organization (WTO) pose significant constraints to the enactment of some climate-friendly trade measures. For instance, most-favored nation (MFN) and national treatment principles dictate that countries cannot discriminate between 'like' products based on their method of production, which includes carbon emissions (Mavroidis & Neven, 2016). Therefore, efforts to modify or reinterpret these principles, to accommodate climate considerations, will be critical moving forward. Ultimately, the success of international trade as a tool for promoting climate action will depend on finding a balance between different nations' rights to economic development and the necessity to protect our shared environment.

Developing Versus Developed Nations play contrasting roles in the politics of climate change. From an economic perspective, developed nations have higher per capita carbon emissions due to an advanced industrial sector as compared to developing nations (Raupach et al., 2007). There's a vast disparity in emissions—developed nations contribute much more towards climate change yet, paradoxically, developing nations are often the ones to be hit hardest by the consequences. This disparity also exists in the ability to negotiate in forums like the United Nations Framework Convention on Climate Change (UNFCCC), where developed nations hold more bargaining power.

Furthermore, the historical responsibilities of developed nations towards climate change often become points of contention in international negotiations. Developed nations, with their established economies, have a longer history of greenhouse gas emissions when compared to developing nations—nations that are now attempting to grow economically (Aldy et al., 2016). Some argue that developed nations should take the lead in mitigating climate change due to their historical contributions to the greenhouse gases already in the atmosphere, while others maintain that all nations, regardless of development status, share equal responsibility in addressing climate change.

Last but not least, there's a considerable disparity in technological capabilities between developed and developing nations. Developed nations often have better access to green technology, which can be used to mitigate environmental damage commensurate to economic growth (Dechezleprêtre et al., 2011). However, developing nations struggle to adopt these technologies due to financial constraints. In conclusion, political negotiations concerning climate change are profoundly influenced by the economic standing and technological abilities of nations, making the distinction between developing and developed nations a critical factor in climate policy.

Chapter 4:
Lobbyists and Climate Policy

In this chapter, we delve into the intricate role that lobbyists play in climate policy formulation, particularly highlighting how lobbying power has a significant impact on the trajectory of climate action. Lobby groups, predominantly backed by major corporations, oftentimes shape climate policy discourse through strategic influence over policymakers (Boussalis and Coan, 2016). Despite their indispensable contributions to policy engineering, these lobbyists can pose formidable impediments to progressive climate action, especially when they espouse the interest of the carbon-intensive industries that frequently resist effective but stringent regulatory standards (Drutman, 2015). However, the power of lobby groups isn't unassailable. The increasing advocacy for climate action from the public, bolstered by emerging lobby groups concerned with environmental preservation, is gradually countering the traditionally dominant corporate narrative (Miller, 2019). Additionally, efforts are now being made to enforce regulations on lobby groups to further assure a balanced and fair influence over policy regulations for the benefit of the environment.

The Power of Lobby Groups

Digging deeper into the role of lobbyists in shaping climate policy, it becomes clear that lobby groups wield tremendous power. Lobbying allows various interest groups, including corporations, advocacy organizations, and others, to exert influence over lawmakers (Drutman, 2015). Akin to a pulling force, the power of lobby groups can tip the scale in their favor, often at the expense of environmental integrity.

How do they manage to exert such influence? Knowledge and finances are key instruments lobby groups use. Through intimate knowledge of legislative processes and policy-making, they posses the ability to sidestep roadblocks most would find insurmountable (Drutman, 2015). They then employ financial resources to fund political campaigns, research, and grassroots movements, further extending their influence.

The financial power of these groups alone can be profound. The fossil fuel industry, for instance, spends millions of dollars annually on lobbying efforts. These finances are used to promote regulatory changes that favor industry growth, often at the cost of environmental protection (Brulle, 2018). By providing significant funding and creating jobs, these corporate lobbies create a sphere of influence that is difficult for policymakers to disregard.

Group interests also play a pivotal role. As collective entities, lobby groups can consolidate and represent the interests of a particular industry, organization or demographic group. For instance, agribusiness lobbies have played a crucial role in promoting policies that prioritize agricultural interests over environmental protection (Lukacs, 2017). Their voice often drowns out those advocating for more sustainable and climate-friendly practices.

Economically heavy sectors such as fosil fuels and agriculture often possess stronger counteracting lobby groups that overshadow voices from environmental lobby groups. This results in a skewed power balance (Brulle, 2018). There's also a tendency for policymakers to adopt strategies and attitudes in sync with the lobbyists that support them, compounding the influence of these powerful groups.

This isn't to say that lobbyists are inherently in opposition to climate policy. On the contrary, there are also profuse lobby groups that advocate for the environment. Renewable energy companies, environmental NGOs, and other groups work collectively to counterbalance the pressure from the pro-fossil fuel lobbyists (Lukacs, 2017). However, these groups often lack the financial power and political clout their challengers have cultivated over the years.

The presence of contrarily influential lobby groups creates a complex, dynamic situation. Decision-making becomes a tug-of-war between lobbyists with compelling, yet often conflicting, interests. Consequently, it's seldom that policies solely reflect the best interests of the environment (Drutman, 2015).

Ironically, the very tools that can be used to protect the climate and invest in sustainable measures, such as policy changes and corporate responsibility, can also be exploited to preserve profit-driven activities that hinder climate action efforts (Brulle, 2018). This irony underscores the vast power of lobbying groups and their lasting, tangible impact on climate policy.

Through campaign contributions, extensive networks, and information control, lobby groups play a major role in shaping climate policy in ways that greatly prioritize their interests. It's not necessarily about being 'for' or 'against' climate action so much as promoting their self-interests, which at times may inhibit environmental progress. Meanwhile, although green lobby groups attempt to counterbalance this influence, they struggle to compete with the financial and political might of their counterparts.

It's important to add that the role of lobby groups can also be deeply influenced by larger socio-political landscapes. Conversely, their actions can in turn affect these very landscapes. Changes in government, public opinion, and international cooperation can either strengthen or weaken the sway of different lobby groups. These groups must, therefore, adapt their strategies according to the changing political tide and devise tactics that enable them to maintain or increase their influence (Lukacs, 2017).

Understanding the power of lobby groups gives us a clearer perspective of the intricate workings of climate policy-making. The complexities involved can, at times, be discouraging. Ultimately, however, it serves to highlight the need for a more transparent, balanced approach to lobby involvement in climate policy. A shift toward dialogue that takes into account the broader environmental and socio-economic impacts, rather than solely focusing on narrow, sector-specific interests, is perhaps needed to address this impasse.

The power of lobby groups, therefore, cannot be underestimated in the context of climate policy. While they can hinder the progress of much- needed legislation and reforms, they also have the potential to promote massive improvements if oriented in the right direction. Their influence is a testament to the fact that in order to make strides in mitigating climate change, it's essential to navigate these powerful intersections of politics and economic interests.

How Lobbyists Influence Climate Policy finds its bread and butter in the process of persuasion; lobbying groups aim to shape and influence policies that in most cases will support the interests of the firms they represent. In the context of climate change, the influence of lobbyists is usually seen in sectors like fossil fuels, manufacturing, and agriculture that might face considerable economic impacts from aggressive climate policies (McMahon, 2020).

One key strategy employed by these lobbyists is the spreading of dubious scientific information to create doubt around the severity of climate change. According to Oreskes and Conway (2010), some lobbyists are known to purposely disseminate disinformation and advocate uncertain scientific arguments that question the validity of climate change, delaying the implementation of robust climate policies. By exaggerating the economic costs and downplaying the environmental benefits of such policies, they aim to dampen public and political will for transformative climate actions.

Beyond mere persuasion, lobbyists also pursue more direct avenues to influence climate policy - campaign financing. Essentially, they donate substantial amounts to political campaigns, often targeting candidates who oppose stringent environmental regulations, thus ensuring political favors in return. According to Brulle (2018) over the past two decades, lobbying expenditure in the US geared towards climate delay exceed the spending by proponents of climate action by a factor of 10:1. The effect of this disproportionate financial aid is a stark delay and inaction on pressing climate issues in policy circles. These methods illustrate how lobbyists can exert significant influence over climate policies, often to the detriment of the global environment.

Big Businesses and Climate Policy lie at a critical juncture where financial interests meet environmental concerns. It's not uncommon to find that many of the world's top polluters are multinational corporations. These companies, driven by profit maximization, may exhibit behaviors that contradict climate change mitigation efforts (Heede, 2014).

Often, big businesses have tremendous lobbying power, which they use to shape climate policies in their favor. In some instances, corporate interests supersede environmental concerns, causing a shift towards policies that are less stringent in reducing emissions (Brulle, 2014). Big businesses have also been known to finance political campaigns, thereby potentially influencing policymakers to design climate policies that favor their operations (Kaplan & Riquier, 2015). This clearly underlines the ability of large companies to sway climate policy decisions and can potentially represent the needs of the business sector over those of the environment.

However, not all big businesses are climate change antagonists. There is a rising trend of companies integrating sustainability into their operations. Some corporations now realize that climate change poses a real threat to their existence, and have begun adopting more sustainable practices. This progressive shift is driven, in part, by consumer demand, as more customers are considering corporations' environmental footprints before making purchasing decisions (PWC, 2020).

Still, the bottom line remains that big business has a crucial role in climate policy. Considering their significant contribution to global carbon emissions, businesses that efficiently manage their environmental impact can substantially influence climate change reduction (FridaysforFuture, 2020). However, to bring about this shift, there is a definitive need for stringent regulations, in tandem with strategic business leadership that considers the interest of the planet without losing sight of profit.

In conclusion, the relationship between big business and climate policy is complex and must be managed with careful balance. Policymakers must work tirelessly to ensure that corporate interests do not dominate the needs of our planet. Business leaders, too, need to step up and adapt to a changing world where profit no longer takes precedence over the health of our planet (Harvey, 2020).

Addressing Lobbyist Influence

Understanding how to address the influence of lobbyists in climate policy is a complex issue requiring multifaceted solutions. Countering the power of these groups involves several strategies. Those strategies include strengthening advocacy efforts for climate action, enforcing regulations for lobby groups, and promoting greater transparency and accountability.

Lobbies, particularly those associated with major industries, have long been among the dominant influences on climate policies. The effect of their influence in many instances has been a slowing or full halt of policies that would aggressively tackle climate change (Hogan, 2020). The rationale behind this influence isn't nefarious, but driven by an economic focus. Many lobby groups represent organizations that could be economically disadvantaged by certain climate change policies.

However, the global climate crisis necessitates that we reconsider this paradigm. The potential damage stemming from unaddressed environmental issues may far outweigh short-term economic interests. The urgency of the situation demands a reassessment of priorities and power structures in the political landscape (Chakraborty & Duke, 2020).

One way to address lobbyist influence is to empower and support advocacy groups dedicated to environmental and climate action. These organizations aim to counterbalance the influence of industry lobbyists by providing lawmakers with research, public opinion data, and compelling arguments for climate-focused legislation (Hojnacki et al., 2015).

Alongside lobbying for institutional change, these advocacy groups also often engage in public education campaigns to garner popular support for their causes.

While the funding and resources of these advocacy groups may be dwarfed by their industrial counterparts, the strength and passion of their constituents can still wield significant power. Public sentiment, fuelled by growing awareness and concern over climate change, can be a potent motivator for politicians (Carmichael et al., 2017). Strengthening these advocacy groups therefore involves amping up their visibility and reach, enhancing their credibility, and increasing public support for their endeavors.

Campaign finance reform is another strategy to mitigate the power of lobbyists. Placing stricter limits on the amount of money that corporations and individuals can contribute to political campaigns would level the playing field. This would enable legislators to make decisions with the interests of the public—and the health of our planet—in mind, rather than those of their biggest donors (Maggetti & Karanikolos, 2022).

Another crucial strategy in addressing lobbyist influence on climate policy is the implementation and enforcement of stricter regulations for lobby groups themselves. Clearer rules for political lobbying, including constraints on revolving-door appointments and gifts to lawmakers would siphon off some of the power that industry lobbyists currently wield (Brulle, 2018).

Moreover, improving transparency in lobbying activities can help shift the balance of power. Mandating stricter disclosure requirements about who lobbyists meet, what policy areas they are influencing, and how much they spend can help to bring the scale of their influence into the public eye (Boyce, 2019). With increased transparency, both lawmakers and the public can be better informed about the depths of lobbyist involvement in the decision-making process.

Implementing such changes would certainly not be an easy task. Lobbyists wield significant influence in part because they are deeply ingrained in the political system. Yet, recognizing this fact doesn't diminish the necessity of lobbying reform (Maggetti & Karanikolos, 2022). We are currently in a state of climate emergency, and the influence of lobby groups can no longer be seen as a mere inconvenient truth but a significant barrier to implementing necessary policies.

Pivotal to this discussion is legislative action. Despite resistance and inherent difficulty, new legislation can provide a stronger framework for more ethical and transparent lobbying activities. Policymakers who understand the urgent need for climate action can lead the charge by introducing and supporting these reforms (Chakraborty & Duke, 2020).

Finally, fostering public awareness and active participation in climate- related decision-making processes is crucial. Grassroots movements and individuals have a critical part to play. Vocal public support for climate change mitigation and adaptation policies can be a compelling force driving legislative changes (Carmichael et al., 2017).

In conclusion, challenging and addressing the influence of fossil fuel and industry lobbyists on climate policy is a daunting task. It is, however, one that we must face head-on for the sake of our planet's future. Bolstering advocacy for climate action, regulating lobbyist activities, and shining a light on the extent of their influence could help turn the tide in favor of environmental sustainability.

Through collective engagement, education, communication, and action, we can begin to reduce lobbyist influence and start creating climate policies that may ensure our world is a livable home for future generations.

Advocacy for Climate Action plays a vital role in mitigating and reversing the detrimental effects of human-induced climate change. The advocacy can take multiple forms, embracing activities like lobbying policymakers, participating in direct actions, raising public awareness, and contributing to local grassroots movements. Advocacy has historically served as a significant catalyst for change, encouraging improvements in the laws, regulations, and policies that govern climate- related actions at the local, national, and global levels (Hadden, 2015).

Effective advocacy initiates dialogue, fosters communication, and transcends geographic boundaries while advocating for sustainable climate policies. For instance, international campaign groups like Greenpeace and Friends of the Earth have been at the forefront of climate advocacy- scheduling meetings with politicians, organizing protests, and leveraging social media to garner mass support for sustainable practices. It's also essential to highlight that numerous smaller community-based groups and activists within academia and other sectors have made commendable strides in shifting the political narrative towards acknowledging and prioritizing the looming climate crisis (Stephan, 2019).

Despite the power of advocacy, challenges still persist. These challenges often emerge from the extent of corporate influence in politics and policies, diverting attention from the gravity of climate issues, and prioritizing short-term economic considerations over long-term environmental welfare. Overcoming these obstacles necessitates a collective, persistent, and vigorous advocacy approach, pushing for a political climate that prioritizes global climate solutions above self and economic interests. Achieving substantial climate action will ultimately require meticulous planning, strategic stepping, and the effective amalgamation of global resistance movements (Newell, 2012).

Regulation of Lobby Groups necessitates an understanding of their immense influence in shaping climate change policy. The need to regulate arises due to the potential for conflicts of interest where lobby groups working on behalf of certain industries might prioritize short-term economic gains above long-term environmental well-being. Regulation efforts require strict transparency laws, clearly outlining the responsibilities and limitations of these groups, alongside monitoring to ensure compliance (Thomas, R., de Figueiredo, J. M., & Richter, B. K. 2020).

The challenges posed by lobbying activities are immense and often not straightforward, which makes robust regulatory mechanisms crucial. An effective regulatory system includes stringent campaign finance laws limiting the amount of money these groups can donate to political campaigns. It also includes implementing policies that ensure that the findings of these lobby groups are impartial and scientifically accurate. Clarifying the blurry lines between lobbying and policy-making roles, and doing away with revolving door politics can also minimize potential regulatory loopholes (Dür, A., & Mateo, G. 2020).

The importance of these regulations lies not only in preventing the potential manipulation of climate policy by vested interests but also in fostering a culture of accountability amongst lobby groups. By having a clear framework in place, it becomes possible to narrow the divide between public interest and private gain, while ensuring that all parties, including lobby groups, contribute constructively towards necessary climate action. However, enforcing such regulations remains a daunting task, and requires collective commitment from policy makers, lobby groups and the public at large (Hoffman, A. J., & Jennings, P. D. 2015).

Chapter 5:
Public Opinion and Climate Policy

The influence of public opinion on climate policy cannot be understated. People have the power to shape climate policy by voting and participating in democracy, thus their perception of climate change is of paramount importance. However, studies suggest that public perception of climate change can oftentimes be swayed by political polarization or lack thereof (Borick & Rabe, 2019). These politicized perspectives prompt a shift in discourse on climate change, shifting the focus from a global environmental issue to a bipartisan debate. Therefore, the role of media and grassroots movements in shaping the public's perception of climate change is critical (Houston et al., 2019). Media representation skews the climate change narrative either by underreporting or sensationalizing it. Grassroots movements, on the other hand, put the power back into the hands of citizens by urging collective action and influencing policy on local, regional, or even national levels. In essence, these movements balance out political bias and empower citizens to take climate action into their own hands (Neumayer, 2019).

Perception of Climate Change

Understanding the general public's perception of climate change is instrumental to appreciate the influence it can exert on climate policy decisions (Weber, 2016). Public attitudes, to a large extent, shape the political landscape regarding climate change which in return influences policy outcomes. So, what is the general public's opinion of the escalating environmental crisis?

Research suggests that the majority of the global populace acknowledges that climate change is a real, urgent, and mainly human-induced problem (Leiserowitz et al., 2015). This consensus has been growing steadily over the years, with an increasing number of people recognizing the paramountcy of quick action. However, this admission varies significantly across regions, nations, cultural and social contexts, as well as demographics.

People residing in nations most vulnerable to the dire consequences of climate change, such as several Pacific island nations, Bangladesh, and parts of Sub-Saharan Africa, typically have high levels of awareness and concern. Similarly, Europeans are generally more worried about climate change in comparison to other western populations. Conversely, the citizens of some developed countries, including sizable portions of the United States, exhibit comparatively lower levels of climate change concern (Lee et al., 2015).

Age also appears to play a role in climate change perceptions. Research has indicated that millennials typically express greater concern for the environment and are more likely to believe that the crisis is human- induced (McCright, 2018). With every rising generation, there appears to be an incremental increase in concern and understanding of environmental issues.

Education and political ideology have been identified as other significant factors influencing perceptions about climate change. Higher education often correlates with an amplified understanding of scientific concepts, including the fundamentals of climate change. Meanwhile, political ideology profoundly affects individuals' perceptions of climate change in several nations (Dunlap, 2016).

Moving on to the political polarization of climate change perception, in certain parts of the world, climate change has unfortunately morphed into a politically divisive topic. This is most evident in the United States, where climate change opinion shows a stark division along partisan lines.

Conservative Republicans are far less likely to believe that climate change is happening and if it is, they contend that it's natural, while the liberal Democrats uphold that it's largely human-induced (Hart & Nisbet, 2012).

These differing perceptions have widespread implications, as they shape not only voters' preferences but also the stance of political leaders. As climate change has become a component of party manifestos and agendas, political leaders' attitudes reflect the ideologies of their respective electorates and vice versa, amplifying the existing polarization.

Political polarization can hamper the passing of bipartisan climate legislation, as was witnessed during the Obama administration. Despite navigating steep political opposition, the Environmental Protection Agency (EPA) established the Clean Power Plan in 2015, which was subsequently rolled back by the succeeding Trump administration; the political divide over environmental issues proved to be detrimental (Hultman et al., 2018).

This division of climate change perception not only influences domestic policy but also international commitments. The critically acclaimed Paris Agreement, for instance, has seen the United States' stance change significantly due to changes in administration and the political polarization on climate change issues (Hultman et al., 2018).

Other nations, too, exhibit a political divide on climate change, albeit relatively less pronounced. In Australia, the Labor Party is more likely to push for a robust climate policy compared to the Coalition, once again with public support reflecting these partisan lines (Tranter, 2011).

In conclusion, the perception of climate change among the general populace is a complex panorama shaped by several social, political, demographic, and cultural factors. These perceptions engender significant repercussions on climate policy and action. Addressing these variances, acknowledging the diverse opinion landscape and seeking middle ground will be crucial in our ongoing pursuit for effective climate policies and practices.

The General Public's Perception of climate change is complex, varying significantly across socio-demographic groups and influenced by numerous factors including political ideology, education, media consumption, and personal experience with extreme weather events (Pidgeon, 2012). A paradox presents itself as the enormity of the climate crisis, underscored by a scientific consensus, doesn't always coincide with the level of importance accorded to it by the public.

While public awareness of climate change is high globally, perception of its severity varies. Some view it as a distant concern, posing threats to future generations or geographically distant communities, but not directly affecting their own lives (Leiserowitz, 2006). This sense of detachment can often result in a passive response to climate change, feeding a cycle of inaction. Others, particularly those who have experienced its direct effects, like flooding or wildfires, regard it as a pressing issue requiring immediate action.

Political ideology plays a marked role in shaping the public's perception of climate change. In countries like the United States, for example, political affiliations can polarize perceptions on this issue. A study by McCright and Dunlap (2011) reported that conservatives are generally more skeptical about the reality of human-induced climate change, whereas liberals are typically more accepting of the scientific consensus. This ideologically-driven divergence in perceptions has significant implications for climate policy-making.

Moreover, the media acts as a bridge between the scientific community and the public, shaping climate change perceptions. The ways in which media outlets portray the crisis, whether they present an uncertain scientific consensus or whether they frame it as a political issue, can significantly affect public understanding and action. In recent years, the surge of misinformation and "fake news" online has also played a damaging role in skewing perceptions and increasing polarization (Hmielowski, Feldman, Myers, Leiserowitz, & Maibach, 2014).

Given the above, the challenge lies in converting public awareness into action. Acknowledgement of the crisis alone is insufficient; rather, an understanding of its severity and urgency is crucial in driving significant and collective action. As such, engaging and educating the public, debunking oppressive climate myths, and presenting a clear conveyance of the facts are vital endeavors. This will not only heighten recognition of the climate crisis but will also underpin the collective willpower necessary to combat it.

Political Polarization of Climate Perceptions introduces a larger issue hindering climate change action - the politicization of this global crisis. Political ideologies shape how individuals cognitively process information. Unfortunately, these ideological constructs dramatically influence people's attitudes towards climate change (McCright and Dunlap, 2011).

Study data affirms a notable divide on climate change opinion across political lines. Identified as the 'conservative white male effect,' conservative white males in the United States are more likely to deny the existence and severity of global warming, adding further evidence to the growing body of literature demonstrating that climate change denial is partially driven by conservative-leaning political ideology (McCright and Dunlap, 2011). However, this phenomenon is certainly not limited to the United States. Political polarization surrounding climate change perceptions has been observed in other democracies such as Australia, Canada, and the United Kingdom (Hobson and Niemeyer, 2013).

While the science of climate change should rightfully be non-partisan, the sociopolitical landscape tends to blur the lines. Polarization over climate perceptions inhibits meaningful dialogue about strategies for mitigating and adapting to climate change. To effectively address these challenges, it is fundamentally essential to recognize and understand the profound effect of political ideology on shaping public perception of climate science (Hobson and Niemeyer, 2013). Comprehensive policy responses require not only understanding the science behind climate change but also accounting for the societal and political factors that shape public opinion.

Influencing Public Opinion

Public opinion on climate change has been subjected to constant persuasion and manipulation. A myriad of entities, mostly outside the conventional political realm, have been intrinsic in influencing how the public perceives the climate crisis and climate policy (Brulle, Carmichael, & Jenkins, 2012).

The media, for instance, plays an extraordinary role in sculpting public opinion toward climate change. It is the principal channel through which the public gets information regarding the climate crisis (Boykoff & Boykoff, 2004). However, the media can also be a vehicle for misinformation and bias, often offering a polarized representation of climate science, thereby swaying public opinion.

The media has a significant power to direct public attention towards or away from the issue at hand, a phenomenon known as 'agenda-setting'. In relation to climate change, some media outlets grant the issue more coverage than others, thus determining how much attention the public pays to the crisis (McCombs & Shaw, 1972). This can in turn affect the public's perceived urgency of the issue, and hence their support for climate policy.

Often, the media also falls prey to 'false balance'— giving equal weight to both scientific consensus and climate scepticism — creating a misleading representation of the climate crisis (Koehler, 2016). This misleads the public into thinking there is still significant controversy within the scientific community regarding human-driven climate change.

Another aspect in shaping public opinion towards climate policy involves grassroots movements. These are on-the-ground initiatives that mobilize communities to instigate societal and political change. A popular example of such movements is the School Strike for Climate, inspired by Greta Thunberg, which has resulted in millions of young people around the world demanding climate action from their governments.

Grassroots movements often rely on personal narratives and local experiences of climate impacts, which can be more persuasive compared to scientific jargon (Moser, 2007). Further, these movements increase civic participation in climate policy which can ultimately lead to stronger and more democratic climate policies (Aylett, 2010).

However, public participation in these movements is often uneven, which can marginalize some voices. Despite their potential to garner public support and influence policy decisions, grassroots movements face significant challenges such as limited resources, local opposition, and state censorship.

Massive advertising campaigns, often deployed by fossil fuel companies and environmental organizations, also shape public opinion. These campaigns utilize the power of imagery, emotion, and rhetoric to connect with people on a personal level. They can either heighten fear and urgency about the climate threat or downplay it, leading the public in the direction they are designed for.

However, its success largely depends on the credibility of the source, which can be a double-edged sword. While climate advocacy groups may appeal to the environmentally conscious, advertisement campaigns by fossil fuel companies might promote skeptic narratives among consumers, thereby diluting concern for the environment.

Interestingly, influencing public opinion is not only a top-down process but also a bottom-up one. Public opinion can shape climate policy through feedback mechanisms (Pidgeon & Fischhoff, 2011). Politicians often rely on public sentiment to drive their policy decisions and platforms. A notable example of this is the Green New Deal, a set of proposed economic stimulus packages in the United States that aim to address climate change and economic inequality, with its popularity shaping the policy discussions in U.S. politics.

In conclusion, the influence on public opinion is multi-faceted, prevalent, and dynamic, impacting our perception of the climate crisis and our role in the solution. As climate change remains a pressing issue, constant vigilance in understanding and challenging the forces that shape public opinion is essential to ensure robust climate policies and the collective will to implement them.

The Role of Media holds a key position in the realm of climate policy, with the potential to both inform the public on climate change, and shape their opinions on the subject. Large media outlets and social media platforms alike play a critical role in public discussions about climate change and the policies needed to address it. With a wide reach and influence, media effectively amplifies scientific findings, spots policy options, and embraces public sentiment towards climate change (Hansen, 2011).

However, the media's role is not devoid of controversy. The two-sided debate typically seen in media outlets tends to present a distorted representation of the overwhelming scientific consensus on human-induced climate change (Boykoff, 2007). This so-called "false balance" often leads to public confusion, risking the perception that the existence of climate change is still up for debate, rather than focusing on solutions and policies to mitigate its impacts. This can play into the hands of political actors who find it convenient to downplay the intensity of the crisis, subsequently stalling transformative policies.

Furthermore, the media can also significantly influence public understanding of the severity and immediacy of climate change. Misleading headlines, lack of contextual information, and infrequent coverage of the issue can contribute to a lack of public urgency. However, researchers have found evidence of increased public engagement and policy response when media coverage of climate change is amplified and accurately portrayed (Schmidt, Ivanova, & Schäfer, 2013). In this era dominated by information and data, media has the potential to create a more informed and engaged public when it comes to tackling climate change. mitigating future impact, and promoting sustainable practices.

Grassroots Movements play a crucial role in shaping public opinion towards the urgency of addressing climate change. These movements, powered by communities and individual advocates, have been instrumental in shedding light on environmental issues from a bottom-up perspective. They've successfully heightened awareness and galvanized action, sometimes even resulting in policy changes. For instance, the anti- fracking movement in various states across the United States saw success as a result of grassroots activism (Stockman, 2016).

On the global stage, movements like Fridays for Future and Extinction Rebellion have rapidly gained traction, utilizing peaceful protests and civil disobedience to demand urgent governmental action on climate change. The power of these grassroots movements lies in their ability to mobilise large portions of the public and demand accountability from policymakers (McKibben, 2019). By creating a sense of shared responsibility and urgency, these movements can heavily influence public opinion. They show the power of collective action, revealing that every person, irrespective of their political, economic, or social standing, has a stake in the fight against climate change.

Paula B. Johnson

Despite their impact, grassroots movements face challenges in sustaining momentum and converting public sentiment into targeted policy outcomes. Institutional barriers, lack of resources, and political opposition are common hurdles. Moreover, while these movements can shift public opinion and foster a greater understanding of climate change, they face an uphill battle making tangible policy changes in a deeply politicised landscape. Still, the historical and ongoing contributions of grassroots movements cannot be underestimated. They are a key driving force for taking these complex environmental issues to the forefront of public discourse and political action (McKibben, 2019; Stockman, 2016).

Chapter 6: Case Study: The Paris Agreement

Turning our attention to a pivotal case study, we delve into the intricacies of the Paris Agreement. Designed as an international response to the irreversible impacts of climate change, the Paris Agreement represents a global consensus in recognizing the urgency of the crisis - a unifying commitment towards limiting global warming to well below 2°C, and pushing efforts further to keep it below 1.5°C (United Nations, 2015).

However, its execution has revealed significant geopolitical hurdles inhibiting its success. A glaring example is the withdrawal of the United States in 2017, which sent ripples across international corridors (Bodansky, 2016). This move not only deflated global efforts but also highlighted the role of national politics, and its potential to sway global initiatives. The Agreement also faced significant challenges in addressing responsibilities and commitments of developed vs. developing nations, creating tensions and slowing down progress (Höhne et al., 2017). Hence, the Paris Agreement, while a significant step forward, provides key lessons on the complexities of orchestrating worldwide cooperation on climate policy and exposes the deep-seated political hindrances we need to tackle.

Behind the Paris Agreement

The Paris Agreement is a groundbreaking environmental treaty that aimed to address the global climate crisis on an international scale, yet its creation faced significant political obstacles. Established in 2015 at the United Nations Climate Change Conference, COP 21, the agreement represented the culmination of years of international negotiations aimed at combating climate change (Falkner, 2016).

The Paris Agreement succeeded earlier attempts to create a global commitment to reduce greenhouse gas emissions, such as the Kyoto Protocol of 1997. Unlike the Kyoto Protocol, which imposed legally binding targets only on developed countries, the Paris Agreement encompassed all countries, illustrating a shift in perception about the responsibility for climate action.

The Paris Agreement's main objective was to keep the global temperature increase well below 2 degrees Celsius compared to pre-industrial levels, and to pursue efforts to limit it to 1.5 degrees (UNFCCC, 2015). This required drastic reductions in greenhouse gas emissions, and it suggested urgency in the transition to sustainable energy sources.

A significant political hurdle in creating a universal agreement was the disparity between developed and developing countries. Developing nations argued that wealthier countries had exploited fossil fuels to grow their economies, thereby contributing largely to the accumulation of greenhouse gases in the atmosphere (Hulme, 2016).

To address this, the Paris Agreement proposed Nationally Determined Contributions (NDCs). NDCs were self-administered plans that countries submitted detailing how they intended to reduce their greenhouse gas emissions. These submissions were not legally binding, but the expectation was that international pressure would hold countries accountable to their commitments (Klein, Pellow, & Brehm, 2018).

An innovative feature of the Paris Agreement was the inclusion of a financial assistance scheme for developing countries. The agreement proposed that developed nations contribute $100 billion annually to a fund to aid developing countries in their adaptation and mitigation efforts. This concept was politically sensitive as it implied a shift in the responsibility from a domestic problem to a collectively shared burden (Klein et al., 2018).

The Paris Agreement also addressed the need for adaptation, recognizing that mitigation alone would be insufficient in confronting the impending threats of climate change. This approach considered the varying vulnerabilities and abilities of countries to prepare for and respond to the impacts of climate change.

Despite its ambitious goals, the Paris Agreement has been marked by controversy and political challenges. Critics point out that there are no enforcement mechanisms to ensure that countries adhere to their NDCs, leading to concerns about the overall effectiveness of the agreement (Winkler & Rajamani, 2014).

Domestic politics have further obstructed the path of the Paris Agreement. In the United States, for example, the Trump administration announced in 2017 that they would withdraw from the agreement. The decision was met with widespread condemnation globally, and ignited fears that other countries might also renege their commitments. Notably, the Biden administration has since rejoined the agreement, underlying how domestic politics can unpredictably shape international climate policy (Kormann, 2019).

Despite these challenges, the Paris Agreement is a critical step towards uniting global efforts to combat climate change. It acknowledges the magnitude and urgency of the climate crisis and sets ambitious goals for global greenhouse gas emission reductions.

The Paris Agreement is at its core, an acknowledgment that climate change is a global problem requiring united efforts. Despite its shortcomings and controversies, it has established a platform for countries to work cooperatively towards a sustainable future.

Origins and Goals of the Paris Agreement trace back to its progenitor, the United Nations Framework Convention on Climate Change (UNFCCC), established in 1992. The UNFCCC's intent was to "prevent dangerous anthropogenic interference with the climate system," a dedication that would be echoed later in the Paris Agreement (UN, 1992). The underpinnings of the Paris Agreement were formed throughout early yearly COP (Conference of the Parties) assemblies under the UNFCCC, but its actual emergence came from mounting international pressure for a legally obligatory and globally unified response to the escalating climate crisis (Hare, Stockwell, Flachsland, & Oberthur, 2011).

What precisely would this globally unified response entail? The answer came in 2015, at the 21st Conference of the Parties in Paris, with the primary goal defined as "(Holding) the increase in the global average temperature to well below 2 degrees Celsius above pre-industrial levels and pursuing efforts to limit the temperature increase to 1.5 degrees Celsius above pre-industrial levels" (UNFCCC, 2015). Undeniably ambitious, this target encapsulates the Agreement's primary objective: to combat climate change by mitigating global warming. Furthermore, the Agreement places a substantial emphasis on financial, technological, and capacity-building support for developing countries, acknowledging the global North-South divide within the context of climate change (UNFCCC, 2015).

Equally critical is the Paris Agreement's commitment to "increase the ability to adapt to the adverse impacts of climate change and foster climate resilience" (UNFCCC, 2015). In the face of an undoubtedly warmer future, panning strategies for climate-conscious adaptation and fostering societal resilience are essential. More so, the Agreement's ambitious goal lays down a legally bounded accountability pathway, pushing nations towards urgency and challenging the international community to maintain a level of climate ambition commensurate with the scale of the problem.

The Paris Agreement is a guiding star rather than a magic bullet, setting goals but leaving countries to chart their paths towards them (Hare et al., 2011).

Achievements and Shortcomings have colored our understanding of the complex relationship between politics and climate change. As seen through the lens of the Paris Agreement, the interplay of these two has notable achievements but is fraught with significant shortcomings. The most critical achievement is perhaps the very creation of the Paris Agreement, which marked the first time that nations worldwide gathered under a common cause to combat climate change (United Nations Framework Convention on Climate Change, 2015).

The Paris Agreement is a testament to the power of international political mobilization and cooperation. It proved that despite differences in national interest, economic stance, and perceived threats, nations could unite when faced with a planet-threatening crisis. This unity, demonstrated by the signing of nearly 200 countries, reflects the capability for global unity that, in itself, is a significant political achievement (United Nations, 2015).

Despite these achievements, the implementation and adherence to the Agreement have seen considerable shortcomings. The most prominent of these is the inconsistency of national contributions to the goals outlined in the Agreement. This inconsistency, largely a result of differing national priorities and lack of enforceable mechanisms, gives rise to a significant disparity in efforts put forth by individual countries (Hare, Fekete, Vieweg, & Schaeffer, 2015).

Another striking shortcoming is the withdrawal of the United States from the Agreement in 2017, reversing the country's commitments to reduce emissions. The move, largely politically driven, highlighted the fragility of international cooperation in the face of national politics. It raised considerable doubts about the viability of global efforts and demonstrated the potential of politics to negatively impact climate change efforts (Houser, 2017).

Lastly, while the Paris Agreement is an achievement in global cooperation, it still falls short in its measures to monitor, verify, and enforce compliance. Through these shortcomings, we see a striking example of how politics affect climate change efforts and the urgent need for political commitment to overcome these barriers (Victor, Akimoto, Hassan, Cullenward, & Hepburn, 2017).

Political Hurdles and the Paris Agreement

The Paris Agreement represented a historic leap in the world's commitment to addressing global warming challenges. However, despite its global resonance and ambitious objectives, implementing the Paris Agreement has been fraught with political difficulties on various fronts.

The politics of individual nations often play an integral role in an agreement's implementation, and the Paris Agreement was no exception. One of the most conspicuous of these political hurdles was the U.S. withdrawal from the agreement during Donald Trump's presidency, an act grounded in economic and ideological differences. The withdrawal, beyond symbolizing a major setback in international climate change cooperation, served as a stern reminder of how unpredictable national politics can compromise global collaborative networks (Hafner & Shiffman, 2021).

Trump's decision was partly influenced by the belief that the Paris Agreement was a raw deal for the U.S., which was required to reduce significant amounts of greenhouse gases compared to other countries like China and India. The decision also stemmed from protectionist tendencies inclined towards shielding American industries from international regulations, particularly fossil fuel industries that were seen as negatively impacted by the agreement (Carrington, 2017).

The repercussions of the U.S. withdrawal from the Paris Agreement were manifold. Economically, it distanced America from potentially lucrative green energy markets and investments, many of which have exponential growth potential. It also undermined the U.S.'s credibility as a global leader, especially on environmental issues.

Although the U.S has re-joined the Paris Agreement under the Biden administration, the intermittent withdrawal under Trump underscores the precariousness of relying excessively on politically unstable national governments to lead the fight against climate change (Friedman, 2021).

Beyond the U.S., the Paris Agreement also spotlights the quagmire of global cooperation and consensus-building. While the consensus-based approach of the agreement was hailed as a democratic feat, it also presented unique challenges. Getting 196 countries to agree on binding emission cuts is no small task, given that every country has different short-term and long-term priorities, varying capacities, and unique environmental and economic contexts.

The concept of 'common but differentiated responsibilities' embedded within the agreement is reflective of this challenge. It recognizes the historical responsibility of developed countries in causing climate change while acknowledging the need for developing countries to grow economically. However, balancing this principle with binding emissions cuts has proved contentious, with developed and developing countries often at odds over their respective responsibilities (Rajamani, 2016).

For instance, developing countries argue that if they are to transition to cleaner forms of energy, substantial financial and technical assistance from developed countries is important. But whether, and to what extent, developed countries facilitate such support often becomes a contentious political issue (Abbas & Abbas, 2020).

Furthermore, enforcement of the agreement has been seen as a universally weak point due to the lack of stringent penalties for non-compliance. This flexibility was crucial for gaining widespread approval; strict enforcement could have deterred many nations from signing. However, the absence of a concrete enforcement mechanism has consequently thrown the agreement's effectiveness into question (Bodansky, 2016).

Climate change doesn't recognize national boundaries, making global cooperation imperative. That's the lofty goal of the Paris Agreement – to unite the world in combating climate change. Nevertheless, national politics, international diplomacy, and the pursuit of developmental goals all put enormous pressure on this shared ambition.

In conclusion, while the Paris Agreement represents a collective step toward combating global climate change, it also highlights the complex geopolitical landscape that colors international environmental agreements. Political uncertainties pose a significant challenge to the sustained, coordinated global action necessary to limit global warming to the target levels set out in the agreement. The key takeaway is that the fight against climate change cannot be won by environmentalists alone; it must be a collective effort, comprising diplomatic, economic, and political mobilization on a global scale.

U.S Withdrawal and Impact The United States, as one of the world's leading carbon emitters, plays a paramount role in global climate change process. The withdrawal of the U.S. from the Paris Agreement in 2017 rang alarm bells worldwide (Britannica, 2021). This decision came at a precarious moment when unified global action was, and still is, of the utmost importance.

The withdrawal was predicated by the administration's argument that the agreement was detrimental to the U.S. economy and American workers (White House Archives, 2017). However, this argument failed to take into account the potential economic benefits associated with transitioning to a green economy, such as job creation in clean energy sectors (Bureau of Labor Statistics, 2021).

The message sent by the U.S. withdrawal, one of disregard for global cooperation, was in itself damaging. It reflected the country's political prioritization of short-term economic gains over the long-term sustainability of the planet. This had the potential to undermine global efforts, particularly those of developing nations that often look to developed countries like the U.S. for leadership and support in implementing climate actions (Levin, 2020).

Moreover, the withdrawal put the U.S. in an unfavorable geopolitical position. The country disenfranchised itself from key decision-making processes around global climate change policy. Simultaneously, it ignored the pervasive and borderless nature of climate change; the consequence of which is felt universally, regardless of a nation's participation in international agreements (Levin, 2020).

However, as of February 2021, under a new administration, the U.S. reentered the Paris Agreement (BBC, 2021). The reentry was a beacon of hope for climate activists, but it also signified the susceptibility of major climate change policies to political winds, which can shift with every new administration. Nevertheless, irrespective of the political landscape, the paramount concern should remain dedicated global cooperation to mitigate climate change impacts, for the sake of both present and future generations (Schellnhuber, 2021).

Global Cooperation Challenges Climate change is an issue so broad, no country can combat it single-handedly. It implicates the globe as a whole, thus requiring the commitment and coordination of all nations. However, this global cooperation crucial in mitigating climate change is plagued by myriad challenges (Bulkeley & Newell, 2010).

One of the significant challenges is the diversity in economic and developmental capacities amongst nations. Developed countries have historically contributed considerably greater to greenhouse emissions, whereas developing nations contribute less due to their limited industrial activities. This differential contribution has raised debates on the equity of burden-sharing in reducing emissions. Developing nations argue that developed nations should shoulder a larger share of responsibility, while developed nations are keen on a more universally equal distribution (Roberts & Parks, 2006), causing a major roadblock in formulating international climate policies.

There's also a tendency for countries to prioritize national interests over global ones, especially evident when it comes to economic gains. Every nation aspires to grow economically, and industrial activity is a fundamental driver of that growth. Limiting industrial activities in efforts to reduce emissions often directly impacts the economic growth of a nation. In other words, the short-term fiscal benefits accruing from industrial activities often supersede the long-term global benefit of climate change mitigation (Gilbert et al., 2012). This leads to considerable reluctance in international cooperation towards addressing climate change.

Moreover, political differences across nations create substantial challenges in international cooperation. Agreements and commitments on climate action reflect each country's domestic political climate. Political spectrum, ideology, public sentiment, and economic imperatives play significant roles in shaping a country's approach to climate action (Andonova, 2009). As political ideologies and circumstances vary wildly across the global diplomatic landscape, achieving consensus on climate goals and policies becomes a mammoth task. In conclusion, global cooperation is indispensable in addressing the existential crisis of climate change, but it faces significant hurdles. Understanding these challenges is as crucial for world leaders as the need to overcome them. As we push forward towards a more sustainable world, we must remember that only together can we weather the storm of climate change.

Chapter 7:
Navigating the Political Landscape
for Climate Action

As we delve deeper into the intricacies of climate action and the political dynamics that obstruct or facilitate it, we confront the sobering reality of the hurdles that stand in our paths. Achieving meaningful legislative reforms often comes up against obstacles such as a lack of political will, policy inertia, and insufficient understanding or prioritization of climate risks (Ypi, 2019). On the international level, challenges are similarly complex, inefficiencies or lack of consensus in global climate governance structures, conflicts of interest or interfactional disparities can stymie progress (Hajer et al, 2015). Yet, not all is doom and gloom: we also witness an encouraging amplification of non-governmental organizations and civil society groups playing more influential roles in advocating for climate action (Hadden, 2015). These entities, be they research think tanks, environmental lobby groups, or citizen activists, can help overcome political barriers, through rigorous policy recommendations, relentless pressure on political entities, and by nudging public opinion and demand towards drastic climate action. However, there remains the inescapable fact that navigating the political landscape to engender meaningful action against climate change is a challenging, often Sisyphean task. It requires an embrace of complexity and a dogged pursuit of incremental gains in a landscape punctuated by setbacks, and marked by the continual need for resiliency and adaptation.

Paula B. Johnson

Barriers to Effective Climate Policies

The onset of climate change has brought about an inevitable need for effective climate policies aimed at mitigating the adverse environmental impacts. However, various obstacles play a pivotal role in hindering the implementation and enforcement of such policies. Whether on a national or international scale, these barriers extend from legislative hurdles, vested interest groups, to international relations, which greatly undermine the fight against climate change.

One predominant barrier lies in the legislative sphere. The formation of climate policies often involves a complex network of government agencies, policy-makers, and legislative bodies. Dysfunctionalities within these systems such as bureaucratic red tape, short-term political cycles, and lack of political will to prioritize climate change issues can deter effective climate action (Harris, 2020).

In many instances, climate policy-making is a slow-moving process hindered by bureaucratic delays. These delays not only impede the timely implementation of policies but also run the risk of policies becoming outdated due to the rapidly evolving nature of the climate crisis.

Short-term political cycles present another legislative hurdle. More often than not, politicians are primarily concerned about issues that will secure their re-election in the near term rather than long-term issues such as climate change. Climate policy-making becomes mired in partisan politics, resulting in delayed or watered-down policies (Chen, 2018).

Lack of political will coupled with misunderstandings or denial about the urgency of the climate crisis also pose significant barriers to climate policies. Policymakers may hesitate to introduce robust measures due to fears of political backlash, or they may altogether abandon the climate change agenda due to skepticism about its impacts (Dunlap & McCright, 2011).

Another significant impediment to effective climate policy-making is the powerful influence of vested interest groups. Particularly, the fossil fuel industry and other major polluters have a vested interest in maintaining the status quo and delaying the transition towards renewable energy.

These groups often exercise their influence by funding climate change denial and blocking or watering down climate policies (Brulle, 2018).

Lobby groups extensively use their resources to fund campaigns and politicians that support their interests, making them a formidable roadblock against climate policies. The skewing of public opinion and climate change discourse through sponsored misinformation campaigns also contributes to policy stagnation in the face of a real environmental threat.

On an international scale, policies related to climate change are not exempt from the complexities of international relations. The inherent inconsistencies among nations in terms of economic abilities, views on climate change, and willingness to cooperate make implementing international climate agreements a multifaceted challenge (Bang, Hovi, & Skodvin, 2016).

Developed countries and developing countries often have differing priorities and capacities in addressing climate change. Developed countries may put more emphasis on mitigation efforts, while developing countries, due to their economic limitations, may prioritize adaptation strategies to meet their immediate needs. This disparity can potentially lead to disagreements and prevent the formation of collective strategies to mitigate climate change.

Furthermore, problems within international relations, such as lack of trust, non-compliance, and geopolitical tensions also pose significant challenges. Countries may not trust each other to act upon the agreed commitments in international treaties which can lead to a lack of cooperation in such shared efforts.

Non-compliance to agreed climate targets due to lack of enforcement mechanisms or capacity constraints at the national level is another problem. Geopolitical tensions also play out within the climate policy sphere, where issues related to trade, boundaries, and power dynamics can lead to disagreements and block consensus on climate agreements.

In conclusion, the barriers to effective climate policies are multidimensional, involving a complex web of factors from the national legislative processes, influence of vested interest groups, to international relations. The urgency and magnitude of the climate crisis motivate a pressing need to address these barriers and promote robust, timely, and inclusive climate policies. To tackle the exceptional challenge of climate change, united efforts along with strategic policies that can navigate through the existing barriers are required.

Legislative Hurdles indeed pose a significant challenge in the realm of climate policy making. Attempts to create effective policies to combat climate change are often met with resistance from certain legislative bodies. Kotcher, Myers, Vraga, and Stenhouse (2017) detailed that certain political ideologies, such as a laissez-faire approach to economics, might propagate opposition to climate change laws for fear of unnecessary government intervention or potential economic slowdown. These ideological divides manifest themselves within the legislative branches of governments, where lawmakers' beliefs inevitably influence their voting behaviors.

The complexity of climate science and climate policies often lead to misunderstanding and misinterpretation. This can also act as a legislative hurdle because lawmakers may not fully understand the scientific basis of climate laws (Bolsen & Cook, 2018). This is not merely an issue of comprehension, but also one of interpretation and communication.

Legislators must grasp the fundamental principles of climate science and translate them into agreeable, enforceable legal statutes.

Without a deep and accurate understanding, lawmakers may inadvertently dismiss critical aspects of policy measures or enact ineffective rules.

Furthermore, the legislative process itself is often slow and cumbersome, posing a hindrance to timely and agile climate action (Bernstein, B. J., Lebow, J., & Kahane, D., 2018). The dire warnings from climate scientists about the urgency of the crisis sometimes fall to the wayside due to the slow-moving nature of the legislative process. Bills addressing climate change may be stalled at various stages, from the drafting of the legislation to its discussion in committees and the eventual voting stage.

This bureaucratic slowdown can render immediate action ineffective or obsolete, proving to be detrimental in the fight against climate change.

International Relations Hurdles represent a significant challenge to the effective implementation of climate policies. Global politics operates on a delicate balance of power, interests, and differing obligations. This becomes even more problematic when viewed against the backdrop of climate change, an issue that requires concerted international effort for meaningful action (Harris, 2016).

The application and operations of international relations are embedded with inherent complexities that pose challenges to global climate action. Firstly, differing national goals and economic realities significantly affect the willingness of states to participate or comply with climate agreements. While developed nations might push for stricter emission regulations, developing nations grappling with poverty reduction and economic growth could perceive these mandates as detrimental to their national interests (Karlsson-Vinkhuyzen & McGee, 2013). Secondly, the principle of sovereignty often stands in the way of effective global climate policies. Countries are apt to prioritize their national interests over reducing carbon emissions, making collective policy enforcement difficult.

Furthermore, geopolitical power dynamics significantly influence the implementation of global climate policies. Powerful nations often use their influence to shape climate outcomes that align with their interests, sometimes at the expense of climate-vulnerable nations (Karlsson- Vinkhuyzen & McGee, 2013). In sum, the entanglement of national interests, economic disparities, sovereignty rights, and global power dynamics make international relations a challenging hurdle for global climate action. Despite these hurdles, it remains imperative for diplomatic leaders and policy makers worldwide to align their interests, navigate these challenges, and undertake aggressive climate action for a sustainable future.

Overcoming Political Barriers

The seemingly stubborn political barriers hindering effective climate policies may seem insurmountable, but they are not. Like any challenge, these barriers can be surmounted with focused determination, strategic planning, and a collective willingness to make fundamental changes.

Policy reforms, strong international cooperation, and a greater role given to non-governmental institutions are among the changes needed to facilitate this process.

A critical first step is the reform of our policies. Many existing policies, from national budgets to trade regulations, inadvertently support environmentally harmful practices (Billingsley, 2014). Hence, there is a need for sweeping policy reforms that support industries striving to be environmentally sustainable rather than those contributing to climate change. Reform is a large undertaking, but it can be initiated gradually and regionally, eventually escalating to make a global impact.

Policies must hinge on the principle of sustainability. Legislative adjustments should encourage businesses to switch to sustainable supplies, research, and development practices. Tax incentives and subsidies can be instrumental in promoting eco-friendly industries and discouraging heavy-carbon-emission industries (Metcalf, 2009). Conversely, sanctions must be in place to deter and punish harmful practices.

While reforms at the national level are critical, we cannot underestimate the importance of concerted global efforts to overcome the political barriers to climate action. Climate change knows no national boundaries; its impact sledged consistently across continents. As such, the mitigation strategies need to be equally unanimous. Traditionally, nations have operated in their best interests, which often conflict with the collective needs of the global community. However, climate change forces us to critically analyze this stance. Maintaining the current model could mean risking the very survival of our collective home, Earth (Vogler, 2016).

In negotiations concerning global climate policies, nations should adopt a shared vision of sustainability rather than individual, short-term economic gains. All nations, large and small, developed and developing, must come together to develop and follow-through these strategic plans for addressing climate change. The Paris Agreement was an encouraging start in this direction (Lane, 2017). Although fraught with hurdles, it brought nations together to pledge towards a common climate change objective.

Beyond necessary policy reforms and the articulation of a shared global vision, non-governmental institutions should play a more pivotal role in overcoming political barriers to climate action. With their own distinct advantages, including technical expertise, political influence, and reach to communities, these institutions can help mediate the politically charged dialogue on climate change.

Non-profit organizations, research bodies, and community-led initiatives can help bridge gaps where political tensions inhibit progress in climate policy. With their unique roles, they can raise public awareness, initiate community action, and contribute significantly to policy recommendations – often exhibited with less ideological bias found in governmental structures (Wapner, 2014).

Importantly, these bodies serve as a conduit for educating and engaging the public in climate issues, who remain the most influential force in shaping the political landscape. The utilization of social media and other modern communication resources by these institutions can help raise awareness and applicable knowledge about climate change, building a society more inclined to enforce climate-friendly political and economic decisions (Nisbet & Kotcher, 2009).

Non-governmental institutions also hold leverage in pressurizing businesses to adopt sustainable practices. By utilizing their influence and the power of the public's sentiment, they can bring about more than government legislation alone (Pulver, 2007).

In conclusion, the task of overcoming political barriers to climate action is a daunting but achievable one. The planet's precarious situation calls for substantive policy reforms at a national level, genuine cooperation on a global scale, and an increased role for non-governmental institutions.

Admittedly, none of these are easy changes. They require shifts in perspective, negotiations through entrenched interests, and the foresight to prioritize longer-term sustainability over immediate gain.

However, the threats and adversities brought by climate change demand we rally past these political barriers. The health of our world, the safety of our species, and the legacy we leave behind hang in the balance. We have no option but to unite, make tough decisions, and courageously face the challenge of climate change. Change, while challenging, is not impossible – our survival depends on it.

Necessary Policy Reforms cannot be overstated. As we navigate the urgent need to address climate change, the inevitable role of public policy becomes clear. At both the national and international level, policy reforms are essential for a wide-scale response to the environmental crisis. These reforms must extend beyond mere rhetoric and partisanship, and instead, foster sustainable practices and green economy, while prioritizing the welfare of all citizens and the planet (Harrison, 2018).

The first necessary reform revolves around strengthening and enforcing environmental legislation. As of now, numerous governmental entities have adopted policies aimed at reducing greenhouse gas emissions, conserving natural resources, and promoting clean energy. However, these policies often lack the enforcement mechanisms necessary to ensure compliance (Garrett, 2019). As a result, enhancing the rigor and implementation of existing environmental laws is imperative.

Additionally, legislation should also aim at restructuring the tax system to include carbon taxation, which would disincentivize fossil fuel use, thereby encouraging a transition towards renewable energy (Metcalf, 2019).

Secondly, policy reform needs to tackle the influence of lobby groups on climate policy aggressively. Despite growing public awareness and concern over climate change, large corporations and their associated lobbyists often stymie policy progression, favoring short-term profits over long-term sustainability (Carmichael, 2020). Legislatures must enact strict regulations to limit the disproportionate sway of these entities.

Reforms should also focus on promoting transparency, collaboration, and accountability between all stakeholders in the climate action dialogue.

This includes governments, businesses, scientific communities, and the public. Ultimately, policy reform needs to empower the transition towards a society that balances economic growth with environmental protection (Harrison, 2018).

Role of Non-Governmental Institutions As we delve deeper into the dynamics of climate change politics, it becomes necessary to shed light on the role of non-governmental institutions. These organizations play a crucial role in climate action, often serving as an influential force in policy making and raising public awareness. They range from scientific bodies that generate crucial data about climate change to advocacy groups that push for political action (Wapner, 2014).

Through research, technical expertise, and well-established networks, these institutions can significantly affect the direction and outcome of climate politics. Many renowned international non-governmental organizations (NGOs), such as Greenpeace and the World Wildlife Fund (WWF), exert their influence by lobbying governments and international organizations, organizing global campaigns, and leading grassroots movements.

Moreover, these non-governmental institutions often act as vital communication links between the scientific community and policymakers. They translate complex scientific data into more digestible, actionable information, facilitating the development and implementation of effective climate policies (Rietig, 2014). They also provide a platform for global and local communities, facilitating the peoples' active participation in the climate change discourse, thereby aiding in the democratic process.

Equally important, in the face of political inaction or counter-productive climate policies, NGOs have a capacity to challenge government decisions and hold authorities accountable for their commitments. They often stand as an emotive public voice, advocating for environment- friendly legislation and putting pressure on politicians to prioritize climate issues in their agendas. This activist role remained crucial, especially in situations where governments were unresponsive to science- based evidence of climate change (Betsill & Corell, 2008).

To conclude, non-governmental institutions perform a vital role within the politics of climate change. Despite occasional criticism for overstepping their professional expertise or representing selected interests, their voices have recurrently been instrumental in pushing for climate action. They articulate a global commitment toward climate protection, standing as a key player in transforming climate politics towards more sustainable paths. Indeed, the continued engagement of these institutions is paramount in navigating the intricate climate-political landscape (Fisher, 2010).

Chapter 8:
Green Parties and Climate Politics

As we shift our focus to the particular political organizations invested in climate advocacy, we uncover the genesis and ascendance of Green Parties worldwide. Its roots steeped in the burgeoning environmental awareness of the 1970s and 80s, the Green movement championed a political ideology fundamentally premised on sustainability, ecological balance, and social justice (Kefford et al., 2018). Unfortunately, Green parties have faced varying degrees of success, marred by entrenched political systems resistant to radical shifts of policy direction, and public skepticism tethered to the misperception of Green politics as singularly climate-centric and thus, neglectful of other pressing public issues.

However, it is arguably in their steadfast advocacy for policy shifts toward environmental sustainability where Green parties make their most significant impact. These parties challenge the status quo at local and global scales, pressing for stringent climate change mitigation measures (Carter, 2018), and standing as the political embodiment of urgent aeonian environmentalist dialogue. As they look ahead, Green parties face multiple challenges, notably the urgent need to broaden appeal beyond their traditionally niche voter base and intensifying the discourse on intersectional environmental policies. Ultimately, they play an essential yet convoluted role in the global political will's transition towards serious climate action.

The Emergence of Green Parties

The 1970s marked a watershed moment for climate activism with the establishment of green parties worldwide. These new political entities emerged with a unique focus on environmental protection, sustainable development, and social justice (Frankland, E., Lucardie, P., & Rihoux, B., 2008). Their ideology tapped into a growing public consciousness of our impacts on the environment and the need for a paradigm shift in our approach to economic growth and development.

Green parties first appeared in Tasmania, Australia, under the United Tasmania Group (UTG) banner in 1972. The UTG contested state elections on an eco-centric platform, marking the world's first green party entry into a political race (McCrea, N., & Letcher, D., 2021). However, the German Green Party, founded in 1980, is often cited as the archetype of a successful green party due to its significant influence on national and international climate policy.

Soon after their establishment, green parties started to emerge in different parts of the world. By the mid-1980s, green parties existed in most Western European countries, extending later to Eastern Europe, the Americas, Africa, and Asia. Regional differences have influenced the focus and strategy of these parties, but climate action, sustainable development, and social justice remain central to their ideology (Bomberg, E., 2002).

Environmentalists were at the forefront of the green parties movement, driven by mounting concerns about pollution, habitat loss, and climate disruption. However, the green political vision extends beyond pure environmental considerations. Over time, green parties expanded their platforms to include commitments towards economic equity, social justice, participatory democracy, and peace - forming what is often referred to as Green Politics (Wissenburg, M., & Schlosberg, D., 2002).

Despite their global distribution, green parties often face similar challenges. Developing nations especially find themselves grappling with resource exploitation, social inequality, and limited economic opportunities. Consequently, these parties' agendas often evolve to confront these intertwined issues, recognizing the interconnectedness of social, economic, and environmental sustainability (Bomberg, E., 2012).

Green Politics emerged from a diverse collection of social movements, including feminism, pacifism, and trade unionism. This diverse background fostered an ideology that prizes collaboration, inclusivity, and diversity. Grassroots activism remains central to green parties, as they often advocate for community-led initiatives and decentralized governance models (Diane, P., & Whittaker, D., 1992).

This focus on local solutions extends to green parties' strategies. Rather than relying solely on top-down governmental action, green parties often promote bottom-up initiatives and engage with local communities directly. This strategy helps decentralize power, perpetuates democratic participation, and fosters creative solutions to environmental challenges on the micro-level (Dobson, A., 2000).

Fundamentally, the core tenets of green parties center around a concept known as "ecocentrism." Meaning, these parties recognize the inherent value in nature beyond its use to humans. Ecocentrism guides party platforms, policies, and communication plans, resulting in a focus on sustainable growth, biodiversity conservation, and climate action (Fox, W., 1990).

The emergence of green parties represented a pivotal shift in global politics. For the first time, parties concentrated on the environment as a centerpiece of their political activity. They recognized the inherent links between social, economic, and environmental issues and advocated for integrated solutions to address these problems holistically.

The rise of green parties also signaled a shift in public opinion. As the parties gained traction, it showed that a significant segment of the population was growing disillusioned with traditional politics and desired policies that prioritize environmental protection and social justice.

Despite their relative novelty, green parties have made significant contributions to raising awareness of climate issues and pushing for sustainable policies. However, their path has not been without obstacles. Parts of society are resistant to change, skeptical of climate science, or prioritizing short-term economic gains over long-term environmental sustainability. In these contexts, green parties have had to navigate a delicate balance, often meeting opposition and challenges to their mandate.

Overall, the emergence of green parties instigated a vital conversation about the intersection of politics and the environment. Notably, they have shown that sustainable development and environmental protection are no longer fringe issues. Instead, they are central to modern political discourse.

Going forward, green parties face continued challenges as climate change progresses and resistance to green initiatives remains. However, their emergence has changed the landscape of global politics and demonstrated that confronting the climate crisis requires a novel breed of political engagement, steadfast in its commitment to ecology, justice, and sustainability.

History and Ideology Green parties originated from social movements advocating for environmental conservation, social justice, and grassroots democracy in the late 1970s and early 1980s. Their ideology, broadly known as 'Green politics,' calls for a systemic change of political and economic structures to ensure sustainable development and tackle the existential threat of climate change (Dryzek, Norgaard, & Schlosberg, 2013).

Green parties generally uphold four core principles, known as the 'Four Pillars.' These include ecological wisdom, social justice, grassroots democracy, and non-violence. Ecological wisdom urges society to respect the natural limits of the planet and encourages sustainable living. Social justice refers to equitable distribution of resources and opportunities, and the condemnations of systemic social inequities. Grassroots democracy denotes bottom-up participatory decision-making processes, influenced by and in the interest of local communities. Finally, non-violence endorses peaceful co-existence and conflict resolution strategies and condemns uses of violence (Frankland & Lucardie, 2008).

In addition to the aforementioned pillars, green parties champion a 'post- growth' or 'degrowth' economic philosophy, which challenges the traditional growth-based economic paradigm. They argue that infinite economic growth is impossible on our finite planet and that such growth often exacerbates social inequality and environmental degradation.

Instead, they propose a shift towards an equitable, sustainable economy focused on well-being (D'Alisa, Demaria, & Kallis, 2014).

Over the decades, green parties have experienced varying degrees of political success, with some even managing to form governments. In Germany, for example, the Green Party has been part of the ruling coalition multiple times, driving the country's transition towards renewable energy and away from nuclear power (Overbeck, 2018). In other countries, however, green parties have struggled to convert their ideological appeal into significant electoral gains, often due to structural political barriers and public scepticism towards their unorthodox economic philosophy.

Nonetheless, green parties have played, and continue to play, a crucial role in advocating for ambitious climate policies globally. They challenge the status quo, raise public awareness on environmental issues, and pressure mainstream parties to adopt greener policies. As the impacts of climate change become increasingly apparent, the relevance and influence of green ideologies in shaping climate politics are expected to grow (Rohrschneider & Dalton, 2018).

Achievements and Failures Throughout history, there have been both successful and unsuccessful attempts at effectively addressing climate change through political means. Some policies have indeed instigated positive change, creating a shift towards more ecologically responsible behavior, yet others have unfortunately fallen short.

One notable success is the European Union's Emission Trading System (ETS), introduced in 2005, which uses a market-based approach to reduce greenhouse gas emissions. This cap and trade system allocates emission allowances to companies; those who exceed their cap can purchase extra allowances from companies who have not used theirs. This system has proven effective, encouraging companies to reduce emissions to save costs, and consequently leading to a significant decrease in EU greenhouse gas emissions (European Commission, 2020).

Yet, political failure in addressing climate change is also evident. The United States, for instance, failed to ratify the Kyoto Protocol, an international treaty dedicated to reducing global greenhouse gas emissions. The U.S. was initially one of the 84 countries that signed the treaty in 1998, but later withdrew, citing economic factors and the lack of commitment from developing countries. This decision was a setback for international climate cooperation and underscored the importance of economic and political factors in climate policy decisions (United Nations Framework Convention on Climate Change, 1998). Hence, the path to effective climate policy remains a complex maze, necessitating the navigation of political, economic, and social considerations in the pursuit of global environmental sustainability.

Green Parties and Climate Advocacy

Green parties have long been champions of climate advocacy. They do not merely acknowledge climate change; their politics revolve around acting on it and seeing it as part of an intersectional issue tied to broader socio-economic problems. An informative perspective emerges when examining the scope and scale of their climate advocacy from both local and international standpoints (Brulle, 2014).

Contrary to many mainstream political parties, Green parties hold environmental sustainability as a core principle. Green parties' policies actively advocate for renewable energy expansion, carbon emissions reductions, biodiversity conservation, and a circular economy (Frankland, Lucardie, & Rihoux, 2008). Many Green parties have proven their commitment to such principles in local governance, pushing for sustainable urban design, stringent waste management, and public transportation improvements (Bomberg, 2012).

Green parties also often encompass a strong element of community involvement and grassroots participation. These democratic practices are essential tools in their climate advocacy strategy, aiming to influence policy-making from the ground up (Kitschelt, 1989). For instance, the German Green Party Die Grünen spearheaded local initiatives for renewable energy transition in hundreds of German towns and cities (Morris & Pehnt, 2012).

The rapid action embodied by Green parties at local levels tends to contrast starkly with the often sluggish approach seen at a broader governmental level. As such, they serve as useful case studies of proactive climate policies that could be possible on a larger scale (Dolezal et al., 2014). Moreover, their grassroots methods are an essential part of building a broad base of climate concern among the public, leading to more significant pressure on governments to act.

Internationally, Green parties are regularly seen to take strong stances on climate issues. They often criticize international climate agreements for not being ambitious enough and call for haste and more significant commitments from nations. A case in point is the European Greens, who have consistently championed robust European Union action on the climate front (Thewissen, 2016).

The international advocacy of Green parties also branches into how cross-border and global issues are handled–emphasizing climate justice, the rights of indigenous peoples, and fair trade (Bratt, 2002). These activities advocate for the understanding that the climate crisis isn't localized, but rather a global phenomenon requiring international cooperation, with responsibilities extending beyond borders.

Emphasizing climate education is another prominent aspect of global Green Party advocacy. They continually work to keep climate change in public and political conversations and push for the incorporation of sustainability education in schools. This advocacy forms a vital investment in future generations of leaders who will be tasked with navigating the complexities of the climate crisis (Carter, 2018).

However, it's worth noting that the impact of Green parties is often constrained by institutional and political barriers. Despite being globally interconnected and sharing common values and objectives, their influence on policy varies greatly from country to country, largely dependent on the individual nation's political structures and systems (Dolezal et al., 2014).

In countries with electoral systems favoring two major parties, such as the United States and United Kingdom, Green parties often have a hard time securing representation and having their voices heard. This situation highlights the need for increased pluralism and diversity in political systems, with Green parties demonstrating clear commitment to the urgency and breadth of the climate crisis (Bryner, 2008).

While Green parties have achieved some substantial victories, they also struggle with inherent issues. Some argue that the scale of their ambition can make their goals hard to realize in practice while others claim that their move towards mainstream politics risks diluting or compromising their initial values (Mudde, 2016).

Despite these challenges, however, Green parties play a critical role in climate activism. They provide a valuable source of opposition to business-as-usual politics, critique conservative climate policies, and drive the necessary sense of urgency. The existence and resilience of these parties underline a growing global consensus that we need to move towards a sustainable future (Müller-Rommel & Spoon, 2009).

In conclusion, Green parties represent a significant political force advocating for climate action. While their influence varies considerably, they are often at the forefront of pushing for broader and more aggressive climate policy implementation, demonstrating potential pathways to a more sustainable future. Future prospects hold both opportunities and challenges as they navigate the complex and often contested political landscape of climate action. Continued perseverance, commitment, and expansion at the local, national, and international levels remain paramount for these parties in their ongoing struggle for meaningful climate action (Thewissen, 2016).

Local and International Impact As the green parties have grown in influence over the years, they have been noted for their impact on both local and international scales. At the community level, these parties play a pivotal role in bringing environmental conservation to the forefront of local politics. Many such parties have implemented programs aimed at reducing waste emissions, promoting recycling initiatives, and advocating for local renewable energy sources (Inglehart, 2018). Local elected officials from green parties have also been instrumental in pushing for stricter zoning laws and other policies to protect natural settings from overdevelopment.

On an international level, the impact of green parties is both compelling and complex. Being part of transnational political groups, such as the European Green Party and the Global Greens, these parties amplify their influence on the global stage. For instance, green parties across Europe have been driving forces behind significant EU environmental legislation. Notably, they pushed for ambitious targets in carbon emission reductions within the bloc, which in turn influenced many other nations' climate policies (Carter, 2018).

Moreover, green parties often speak out against environmental injustices worldwide. Many have taken firm stances against exploitative practices such as deforestation, overfishing, and extraction industries, frequently pushing for policies and sanctions against countries promoting such practices. As eco-friendly parties gain more traction, their messages garner wider audiences on the international scale, amplifying the global discourse on environmental conservation (Inglehart, 2018).

However, the impact isn't always positive or unified. The diverse array of green parties worldwide often has different priorities and strategies, which can lead to inconsistent approaches to climate policy internationally. In some instances, the unique local and national contexts in which these parties operate can lead to rollbacks or weakening of environmental laws in the name of economic development or security (Carter, 2018).

In conclusion, amid these challenges, the role of green parties in shaping local and international political landscapes cannot be underestimated.

Their impact, though varied, signals a gradual and necessary shift in the political paradigm towards more sustainable policies and practices. With further growth and unified approaches, green parties can enhance their impact, acting as potent agents of change in climate politics.*(Inglehart, R. (2018).*

Future Prospects and Challenges The surging global awareness of climate change and environmental crisis is fostering a new wave of green political thought and action. However, the road is fraught with challenges that Green parties must overcome to effect meaningful change (Dryzek, Norgaard, & Schlosberg, 2011).

The future growth of green politics hinges heavily on their ability to connect with broader sections of society by addressing economic and social justice issues in conjunction with environmental considerations. Additionally, Green parties must innovate to overcome traditional political dichotomies and build broad-based coalitions working for systemic integrations towards sustainability. This involves mobilizing campaigns that intelligently combine advocacy, electoral, and direct action tactics (Delmas & Young, 2009).

The challenges these Green parties face are manifold, some of them inherent in the changes they seek to implement. Implementing green policies often involves a radical paradigm shift which can be met with resistance from sectors of society that fear potential economic consequences. The dependency of many countries on fossil fuels, for instance, creates a significant barrier to the transition to renewable energy. Furthermore, there is the constant challenge of political opponents downplaying environmental concerns and climate change realities. Overcoming these hurdles requires not just political will, but also educating the public and building social consensus on the urgency and benefits of adopting sustainable practices (Javeline, 2014).

Chapter 9:
The Future of Climate Politics

The future of climate politics hinges largely on the integration of environmental concerns within electoral platforms, sparking three critical conversations, namely elections (Plumer, 2019), legislative reforms, and transformative leadership. Recent political climate actions, like the aggressive renewable energy commitments of Denmark's winning social- democrats and other climate-friendly waves in European parliamentary elections, are clear examples of the increasing weight of climate issues in shaping electoral outcomes (Newell & Paterson, 2018). Anticipated future trends forecast an even more pronounced role of climate dynamics in political landscapes globally. Forced by growing public pressure, governments will inevitably need to adapt their political structures. This adaptation must include legislative reforms, which specifically focus on enshrining climate action as a legally and politically binding duty, following the trail-blazing examples of certain Scandinavian countries.

Scholars also emphasize the critical need for transformative leaders who will courageously champion interventions aimed at drastically reducing carbon emissions despite the potential political implications – leaders who understand the stakes and are willing to take measures that may be unpopular in the short-term but are necessary for the long-term perseverance of our planet (Archer, 2018).

Climate Change and Elections

The discourse on climate change and politics has gradually transgressed from mere theorization to practical manifestations in the reality of elections. The procedure to select political leaders, particularly in democratic nations, has now taken center stage in the fight against the global climate crisis (Loft, Juhola, & Dransfeld, 2020). Elections are a crucial time when societal divisions and opinions on various issues, including climate change, come to the fore.

The issue of climate change is no longer secondary in electoral campaigns. In recent years, it has emerged as a significant deciding factor among voters (Bernauer and McGrath, 2016). There has been an increased voter tendency to assess candidates based on their knowledge of the climate crisis and their proposed policy solutions. This has triggered a trend among political parties to integrate robust climate change agendas into their political manifestos.

This is not to suggest that this increased awareness always results in the election of environmentally conscious leaders. Inconsistent voter behaviors, party polarization, and the overriding economic concerns often influence electoral outcomes, overshadowing the climate change agenda. These limitations, however, do not diminish the power that elections hold to shape the future course of climate politics (Hobson & Niemeyer, 2013).

The entrenched party-policy linkages also play a significant role. Some parties tend to raise climate change issues more often than others, especially green parties, implying that their electoral success can directly impact the process of climate policymaking (Rüdig, 2012). However, the political discourse in elections can be skewed. For instance, parties with traditional ties to carbon-intensive industries often undermine the magnitude of the climate problem in their campaigns, affecting public perception and voting behavior.

Also notable are the electoral politics at the municipal level, which have profound implications for climate change. Local governments, with their increased scope for direct intervention, often form a critical link in implementing and sustaining climate mitigation strategies (Sharifi & Castón Broto, 2016). Hence, the choice of local government leadership through elections becomes a significant determinant of the effectiveness of these measures.

The rise in climate-induced disasters has been another game-changer in electoral politics. These disasters highlight the candid reality of climate change, influencing public sentiment and voting choices (Congleton, 2006). These events present an opportunity for politicians to demonstrate climate leadership, which if harnessed effectively could resonate positively with the voters.

Furthermore, the role of climate activism within the sphere of elections is a considerable force. Climate activists and non-governmental organizations (NGOs) work tirelessly to enhance the visibility of climate change as an electoral issue and push political parties to commit to strong environmental policies (Grumbine, 2017). Through different forms of advocacy and recent youth-led climate protests, these entities seek to affect voters' preferences and spur political actors to action.

Despite the progress in mainstreaming climate change within electoral discourse, there is still a long way to go. Strategies aimed at enhancing climate change knowledge among citizens, integrating climate change into political education, and ensuring campaign financing transparency could potentially strengthen the role of climate change in shaping electoral outcomes (Chen, 2017).

Media portrayal of climate change during elections is another aspect needing attention. Biased or misleading media reports can jeopardize voters' understanding of the issue and thus their voting choices (Brulle, 2014). Therefore, improved climate science communication and quality reporting could support informed voting.

Considering the global urgency of the climate crisis, the potential role of elections in shaping climate politics cannot be understated.

The relationship between climate change and elections calls for rigorous academic exploration and policy interventions to ensure the translation of expressed environmental concerns into electoral choices and subsequent policy actions.

Addressing the climate crisis must go beyond the election cycle to ensure the enforcement and continuity of climate policies. Often, short-term political gains sideline long-term environmental goals. Therefore, it's crucial to establish mechanisms that will ensure climate policy continuity, irrespective of the political season.

In conclusion, climate change and elections form an intricate nexus influencing the direction of climate politics. From shaping public opinion, influencing party manifestos, to determining the commitment of political leaders towards climate action - elections play a vital role.

Observing how this nexus evolves and strategizing political action accordingly is essential for advancing towards a more sustainable future.

Climate Politics in the Election Cycle significantly impacts the trajectory of environmental policy and climate action. During the election cycle, climate change often becomes a polarizing issue, shaping political campaigns and influencing voter turnout. Heller, Chambers, and Fiorino (2014) argue that the politicization of climate change heightens during election seasons, thereby making it critical for environmental policy developments (Heller et al., 2014).

The manner in which politicians command the narrative around climate science can greatly influence public perceptions and voting behavior.

They often water down or amplify the urgency of climate change based on their political affiliation and campaign strategy. Moreover, climate skeptic candidates frequently support policies favoring fossil fuels and other polluting industries, while those acknowledging the gravity of the situation call for swift policy changes in favor of renewable energy and sustainability (Fisher et al., 2012).

Another critical aspect of the election cycle is how campaign donations can steer politicians' stance on environmental policies. There's an observable trend of political candidates receiving funding from industries such as oil and gas, which could potentially skew their position toward climate science denial or leniency on environmental standards (Brulle, 2013). As a result, the policymaking process may become compromised, leading to insufficient climate action during their tenure.

Policies on climate change also often seesaw between administrations, particularly when the incoming government has a different view on climate change than its predecessor. This inconsistency in climate action policies can hinder the continuity and effectiveness of the steps taken towards mitigating climate changes (Lachapelle, 2018).

To conclude, despite the challenges, elections present an opportunity for significant climate action. If citizens push climate change to the forefront of political discourse, they could compel politicians to take clear stances on it. In turn, this could lead to greater transparency and accountability in climate politics and policy, fostering more sustainable approaches to governance long into the future (Capstick et al., 2015).

Predicting Future Trends is an essential aspect of understanding and responding to the political factors that contribute to climate change. As scientific projections show us that global warming is a growing threat to our environment and health, we need to comprehensively analyze the trajectory of political trends to course-correct before it's too late (Intergovernmental Panel on Climate Change, 2014).

The rise of environmentally-focused political parties and candidates across the globe in recent years is an encouraging development. As more citizens become aware of the severity of the climate crisis, there is a trend towards voting for leaders who prioritize sustainable policies and can address the profound implications of climate change (DeSombre, 2020). However, the degree of success these parties will have in fundamentally altering policy trajectory remains to be seen, as entrenched economic interests and ideologies pose significant barriers.

In terms of climate policy, future trends could be strongly influenced by geopolitical shifts. It's expected that debates over who should shoulder the burden of climate change mitigation - developed countries versus developing - will continue to dominate climate negotiations (Victor, 2015). More affluent nations, having historically contributed more to greenhouse gas emissions, may face increasing pressure to lead in emission reductions and to provide technical and financial assistance to less developed nations.

The role of lobbying in shaping climate policy is another trend to watch. While they have thus far mostly advanced the agenda of fossil fuel industries, it's predicted that lobbying groups, recognizing shifting public demand and productive investment opportunities, may increasingly advocate for climate action (Skocpol, 2013). Yet, this change will likely not be without contestation, as existing industry interests continue to exert influence over political decisions.

Finally, climate denial, often politically motivated, has had a major influence on policy-making. The presence of climate skeptics in politics is expected to decline as the evidence of extreme weather events and scientific consensus becomes harder to dismiss (Dunlap & Jacques, 2013). While predicting future trends making assumptions and accepting uncertainty, it's clear that political actions in response to climate change will be pivotal in shaping our shared environmental future.

Adapting Political Structures

The progress of human civilization and climate change are intrinsically linked, and thus, we can't overlook the potent role that politics plays in mediating this relationship. Yet, the political institutions and structures we've relied upon were primarily suited for the challenges of the past and not the existential threat of climate change. Consequently, both our political systems and ideologies must evolve and adapt to this rapidly changing landscape.

Legislative measures form the backbone of any comprehensive political response to climate change. Currently, our legislative frameworks lack the ambition and scope required to address the challenge at hand. A significant paradigm shift is necessary, one that transforms archaic systems and introduces policies which center around sustainability and preservation (Stokes, 2013).

National governments worldwide must enact laws that both mitigate greenhouse gas emissions and facilitate adaptation to unavoidable climate change impacts. Such legislative reforms would ideally include strict emissions standards, incentives for renewable energy and conservation, and resources for communities most affected by climate change (Rabe, 2004).

While important, reactive legislation addressing the damage already inflicted isn't enough. Instead, proactive laws that prevent or minimize further damage are critical. For instance, implementing "green amendments" into constitutions - legal provisions explicitly guaranteeing the right to a clean and healthy environment - could compel governments to prioritize ecological health and sustainability (May & Daly, 2016).

Apart from updating existing laws, innovative pieces of legislation like carbon pricing measures can incentivize cleaner practices and energy use. Essentially, this approach attempts to correct market failures by imposing a tax on the carbon content of fossil fuels, thus promoting less harmful energy sources and consumption choices (Metcalf & Weisbach, 2009).

However, legislation alone cannot provide a panacea for the climate crisis. For any meaningful change to occur, the institutions that create and enforce these laws must also embody the principles of environmental sustainability. Thus, the role of future leaders and decision-makers is crucial (Jacobs, 2012).

Universally, political leaders have a responsibility to lead from the front and guide their electorates out of the climate crisis. However, the leaders directly shaping future climate politics will need a finely-honed understanding of the intersection between socio-economic development and environmental conservation. An unwavering commitment to prioritizing humanity's long-term survival over short-term gains is essential (Hulme, 2009).

Moreover, these future leaders must reflect the diversity of the communities they represent, incorporating voices and perspectives that have historically been marginalized. Indigenous peoples, women, and young people have been disproportionately affected by climate change, yet their voices are underrepresented in decision-making. Greater inclusion of these viewpoints in leadership roles enriches policy-making, lending credence and weight to climate justice claims (Schlosberg & Collins, 2014).

Unquestionably, future leaders will have to navigate a complex web of stakeholders. Each party, from big business to grassroots community groups, has varying interests and influence. To mediate these interests while pursuing aggressive climate actions, leaders must inject transparency into the lobbying and policymaking processes. This includes not only uncovering undue influence but also making public policy decisions more understandable and engaging for ordinary citizens (Fischer, 2019).

In addition, the role of international political structures is of pressing importance. As climate change transcends national boundaries and jurisdictions, international cooperation and commitment are indispensable. Future leaders must prioritize working through multilateral agreements and global institutions to ensure a coordinated, effective response (Mayer, 2016).

Yet, within these dialogues, traditionally less powerful nations must not be overshadowed by more powerful ones. Developing nations, while contributing the least to global greenhouse gas emissions, stand to suffer the most from their impact. This socio-economic disparity must be addressed by empowering these nations in international climate politics (Roberts & Parks, 2007).

Ultimately, political adaptation to the climate crisis encompasses far more than drafting new policies or electing innovative leaders. It requires the reshaping of structures and processes that have guided human civilization for centuries. By embracing adaptability, inclusivity, and justice as core principles, political institutions can better serve the future that awaits us in a dynamically changing climate landscape.

Legislative Reforms for Climate Action play a pivotal role in addressing climate change. In the contemporary arena, the influence of politics is vast in shaping climate action, with significant potential to either augment or hinder efforts to mitigate climate change effects. Legislation is an essential tool in defining environmental policies, and creating and implementing effective laws can result in landmark strides towards a sustainable future.

The dynamic nature of environmental issues necessitates legislative flexibility. One example of this is the shift towards an emphasis on carbon pricing. Economists and climate scientists contend that placing a monetary cost on carbon emissions presents an effective approach to curbing greenhouse gases. This type of policy ensures that companies are held financially accountable for their environmental impact, thereby incentivizing a reduction in their carbon footprint (Scott, Geden, & Quéré, 2021). This kind of legislation, if amplified on a global scale, could represent a significant step forward in climate action.

The cornerstone of any successful legislative reform, however, must be robust accountability measures. Beyond merely creating policies, enforcement is critical. Too frequently, lack of oversight and penalties leads to policies lacking in their execution, undermining their initial intent (Mazmanian & Kraft, 2019). Further, international collaboration in climate policies is also essential. Climate change is not restricted by geographical boundaries and thus demands a collective approach.

Agreements such as the Paris Agreement epitomize the kind of cooperation needed to combat climate change but equally showcase the immense political challenges involved in achieving such consensus (Falkner, 2016).

The Role of Future Leaders is paramount when considering the future of climate politics. Climate change poses an existential threat, and it will be incumbent upon future leaders to comprehensively address this multifaceted challenge. The leaders of tomorrow will require exceptional acumen for legislative nimbleness, resilience in negotiating international relations hurdles, and staunch dedication to implementing necessary policy reforms (Intergovernmental Panel on Climate Change, 2014).

Primarily, future leaders need to bridge the current gap between scientific understanding of climate change and the political will to act. While science provides unmistakable evidence for climate change, the political landscape has traditionally been slow to act. To facilitate more agile policy-making systems, future leaders must uphold scientific intelligence into the legislative process while continuously maintaining a dialogue with the scientific community (National Research Council, 2010).

Moreover, navigating the international relations arena is a core prerequisite for effective climate policies. Future leaders must foster global collaboration and strategic partnerships. Climate change extends beyond national boundaries and demands a holistic, globally mindful perspective. This involves striking a balance between developing and developed nations, ensuring climate policy's equitable implementation (United Nations, 2015).

Notably, necessary policy reforms underline the critical need for future leaders to be brave, innovative, and ethical. We need leaders who can not only introduce new policies that align with environmental science but also overhaul existing structures that block climate action. Additionally, leadership must encourage non-governmental institutions' involvement in climate matters, thereby facilitating comprehensive societal participation in battling climate change (Hale, 2016).

In conclusion, future leaders play a decisive role in shaping the trajectory of climate politics. Their willingness to embrace science, orchestrate positive international relations, and boldly reform policy will significantly impact our ability to mitigate the dire consequences of climate change. Realizing this, it is crucial that individuals, communities, and nations actively support leaders who demonstrate a commitment to these critical attributes and the capacity to navigate the complex landscape of climate politics (Cox, 2015).

Conclusion

Our exploration of the intersection between politics and climate change reveals a poignant yet complex phenomenon. We have uncovered how political institutions, policymaking, public perceptions, economic perspectives, lobbying influences, scientific biases, and international agreements play colossal roles in shaping global climate action—or sadly, our lack of it (Leiserowitz, Maibach, Rosenthal, Kotcher, Ballew, Bergquist & Gustafson, 2020).

The examples drawn across the chapters paint a holistic picture of these effects at play. We have trudged through the challenging terrains of political barriers, witnessed the power struggles of lobbyists, grappled with the mires of public opinion, and stood in the battlefronts of science and climate denial. Our journey through the Paris Agreement has shown us the high stakes of international cooperation in combatting a crisis that recognizes no borders (Hulme, 2018).

We have seen how politics can, unfortunately, obstruct climate action just as it can fuel it. Misuse of power, misrepresentation of facts, and misguided ideologies have often led to policies that negate the efforts to curb global warming (Boykoff, 2019). At other stages, we have witnessed a glimmer of hope; policies that embody environmental stewardship and international corporations committed to sustainable practices show that the scales can tip towards positive change (Hoffman, 2021).

Key Lessons

Several key lessons emerge from the climate-politics interface that we must heed. Firstly, it becomes evident that political will is a catalyst for systemic and impactful Climate Change mitigation. Responsive governments, armed with robust climate policies, can drive notable positive change. Moreover, the weaving political, economic, psychological, ethical, and technical dimensions of climate change call for interdisciplinarity in climate action (Dunlap, 2013).

Secondly, the enigma of climate change exceeds the political domain. It transcends borders and generations, urging for far-seeing and cooperative international solutions like the Paris Agreement. Thirdly, transparency and accountability must underscore the political response to climate change. The role of lobbyists begs the need for a commitment to ethical lobbying that prioritizes the collective good over individual interests (Gifford,2011).

Lastly, it has become apparent the vital role of scientific evidence in guiding climate policies. The politicization of environmental science threatens this fundamental principle, emphasizing the importance of maintaining the autonomy and reliability of climate science (Pearce, Brown, Nerlich & Koteyko, 2015).

Steps for Individuals

Though the problems seem daunting, they are not insurmountable. There are steps individuals can take, starting with awareness and education about climate change and its politics. This knowledge base will empower citizens to vote informedly, encouraging leaders who prioritize environmental stewardship (Schneider, Rosencranz, Mastrandrea & Kuntz-Duriseti, 2010).

Individuals can also play a role in influencing public opinion. By engaging in conversations, creating awareness, and debunking misconceptions, we can shape a society that recognizes the urgency of climate action. Also, supporting green businesses, reducing waste, and mindful consumption are simple yet influential everyday practices that can make a difference (Ballew, et al., 2019).

Above all, remember that every voice matters. Your involvement in grassroots movements can bring about collective action that transforms communities and influences policy (Fisher, 2013).

Strategies for Policymakers

For policymakers, many paths lie ahead. Their primary task is to ensure political structures and legislation embody sustainability, aligning economic, social, and environmental objectives (Naustdalslid, 2011).

Climate policy should be a priority in all industries, not just a few, and elected officials must aim for the long-term benefits, not temporary and short-sighted wins (Giddens, 2011).

Critical in this endeavor are green parties and non-governmental organizations, as their advocacy can accelerate climate action significantly. They can drive innovative climate solutions, help shape public opinion, and lobby for policy reforms favorably (Carter, 2018).

Solutions should also involve active engagement with scientific research in policymaking, incorporating the best available data and predictions.

Scientists and policymakers must work together to develop actionable goals, accounting for both the big picture and the local context (Oreskes, 2004).

In the end, confronting the global climate crisis demands aligning political actions with scientific facts and societal needs. Our judicious understanding of how politics and climate change interact can usher in an era of informed, cooperative, and impactful climate action. As we navigate this precarious interface, we are not just observers. We are agents of transformative change. Our informed decisions, collective actions, and political advocacy can help steer humanity towards a sustainable and resilient future (Wapner, 2010).

Learnings from the Climate-Politics Interface

As we have navigated this complex topic throughout the book, one clear revelation stands out: the intersection of climate and politics is more deeply entangled than one would initially presume. This has placed an inordinate responsibility on political leadership to progress global action against climate change. With this burden, however, comes remarkable opportunities to leverage political mechanisms to promote sustainable development and preservation of our environment (Dietz & Stern, 2002).

One of the first fundamental takeaways from our exploration is the crucial role politics plays in addressing climate change. Political leaders and institutions hold power over the creation, development, and implementation of policies that can profoundly influence our environmental footprint. However, the misuse of environmental science for political agendas has often led to misinformed policies and public denial or indifference towards climate change, thus obstructing effective climate action (Oreskes & Conway, 2010).

We have learned that economic considerations are key driving forces behind climate policies. The affordability debate and impacts on international trade and development have been pivotal in shaping climate laws and regulations. Moreover, while economic costs of climate action have been a focal point for critique, it's equally essential to acknowledge and leverage the economic opportunities this transition offers (Delina, 2018).

The role of lobby groups and big businesses in climate policy formulation provides an interesting dichotomy. On one hand, they may promote industrial interests at the expense of environmental welfare (Grumbach, 2015). However, their influence can be crucial in advocacy for climate action, thus showcasing another opportunity for political will to regulate their influence positively.

Our exploration also delved into public opinion and its influence on climate policy. We understand now how perceptions of climate change are often politically polarized, and therefore the role of media and grassroots movements in shaping these perceptions can't be underestimated. These stakeholders hold the potential to bridge this divide and unite public opinion in favor of climate action (Leiserowitz, 2006).

A significant area of focus was the Paris Agreement, which exemplifies both the achievements and challenges at the global-political level. We learned that international cooperation is key, but it also comes with numerous hurdles, such as the withdrawal of key nations and the issues of fairness and transparency at stake. This aspect has further reiterated the need for global political solidarity and commitment to climate action (Keohane & Oppenheimer, 2016).

The emergence of Green Parties and the importance of political leadership at local, national, and global levels in advocating for climate policies were also explored. Their contributions, though significant, also come with their own set of challenges and future prospects that warrant further attention.

In regards to the future of climate politics, we garnered insights into how climate change could become significant in future elections. Legislative reforms required for climate action and the need for fresh, compassionate, and competent leadership emerged as crucial needs for a future with successful climate policies.

This journey through the climate-politics interface has demonstrated the intricate web that these two spheres weave together. Every aspect of political systems - from elections to policymaking, from local to international politics - has effects on and is affected by climate change (Fisher, 2013).

Thus, our exploration encourages us to view climate change not just as an environmental issue, but as a complex and deeply rooted social, economic, and political challenge. For effective climate action, acknowledgment of this multifaceted nature and a holistic strategic approach to policy-making is required. Each political tool should be utilized purposefully to secure a sustainable and climate-resilient future.

In conclusion, this examination of the climate-politics interface has been insightful and thought-provoking. Strides have been made, but the road to truly integrated and effective climate policy is long. It is abundantly clear, however, that navigating this interplay is not just a possibility but a necessity. With political will, international cooperation, and public support, it is indeed possible to address climate change effectively and comprehensively.

Key Lessons One of the fundamental understandings gained from this exploration of climate change and politics is that the two themes are inextricably linked. Climate change is not merely an environmental crisis; it is greatly influenced by political motives, actions, and power dynamics. From the top echelons of government to the minute details of policy crafting, politics plays a crucial role in the direction and effectiveness of climate action (McGuirk, 2014).

We learned that effective policies addressing climate change often meet resistance not only due to economic constraints but also because of political ideologies or associations with lobbying groups (Wynes et al., 2017). The inducement and power bestowed upon politicians by lobbyists from fossil fuel industries have profound impacts, often leading to denial or downplaying of climate change issues. This shows us that for climate action to be successful, it is imperative to address the conflicting interest that mires the political process (Yeo, 2020).

Another lesson is that public perception and understanding of climate change are crucial for climate policies to be effective. Politics shapes the public's understanding and perception of climate change, often leading to polarization of views (McCright & Dunlap, 2013). This illustrates that it's vital to communicate scientific findings in an accessible and unbiased way to foster climate literacy and inform public opinion.

A review of case studies like the Paris Agreement showcased the difficulty of achieving global cooperation in combating climate change. The withdrawal of the United States emphasized that political instability can greatly affect the world's climate action efforts. This emphasizes the need for climate agreements to be resilient to changes in political landscapes (Victor, 2013).

Last, the future of climate politics is likely to permeate more into election cycles and influence the political structures of governments around the world. Climate action will require political will from the leaders and reform in legislation (Chapman, Lickel, & Markowitz, 2017). The emergence of green parties and climate-focused candidates signal a shift in this direction but also poses its own challenges. These are the key lessons learned about the interplay between politics and climate change. It not only enriches our understanding of the dynamics but also informs strategies to advocate for better climate policies effectively.

Navigating the Future

In analysis of the intricate intersection of climate change and politics, the most pertinent endeavor comes in formulating effective roadmaps for the future. Addressing climate change successfully requires not just political action, but a genuine commitment to transformative policy and cultural changes at all levels - individual, corporate, and governmental.

Individual actions, while seemingly minuscule in the grand scheme, are critical in solving the climate crisis. When considered collectively, lifestyle changes like reducing energy consumption, conserving water, recycling, and mitigating personal carbon footprints can have substantive impacts (Roser & Ritchie, 2020).

Similarly, opting for sustainable transportation, adopting plant-based diets, and actively engaging with and supporting local sustainability initiatives can have a ripple effect across communities. These shifts, however, require public education and awareness to ensure mindfulness of the impacts of our everyday choices, driving a societal tend towards environmental responsibility.

Moreover, citizens have powerful voices that can be used to foster change. By engaging in climate literacy, individuals can gain a comprehensive understanding of climate change and its implications, empowering them to communicate proficiently about the issue, and advocate for effective responses (Cook et al., 2016).

Voting for leaders who acknowledge climate change as a priority further plays into this, alongside active participation in climate protests to demand policy change. An informed and engaged electorate contributes forcefully to a political environment that propels climate action.

On a larger scale, corporations hold immense power to foster climate solutions. Businesses—including those most responsible for greenhouse emissions—must adopt responsible practices. Many corporations have begun to recognize the economic potential of sustainability, responding to investors and consumers who increasingly demand corporate social responsibility.

These include transitioning to renewable energy sources, integrating circular economy models to reduce waste, investing in green technologies, and aligning business models to the United Nation's Sustainable Development Goals (Mayer, 2018). Also, transparency in reporting carbon footprints fosters corporate accountability, pushing businesses to reduce emissions.

Moving the lens to policymakers, a vital first step entails acknowledging climate change as an urgent global crisis that requires immediate action. Such acknowledgment forms the basis from which to create robust, science-based climate policies that address both mitigation and adaptation strategies.

The revitalization of global cooperation to enforce international climate agreements—such as the Paris Agreement—is crucial, particularly from high-emitting countries. This necessitates the detangling of climate action from partisan politics to ensure policy consistency across different political administrations.

While international regulations are essential, national strategies tailored to the unique circumstances of each country must also be given priority. By incentivizing the transition to renewable energy and removing subsidies from fossil fuels, governments can spur the growth of green economies.

Additionally, governments and lawmakers need to address the influence of powerful corporate lobbyists on climate policy - by establishing regulations that prevent policy manipulations and implementing transparency in political funding (Brulle, 2018).

Furthermore, there is a strong need for integrating climate change education into national curricula to cultivate a society that understands and values the environment. By fostering a culture of sustainability, governments can intersect an array of policies—from urban planning to agriculture and transportation—with climate action.

Finally, the integration of climate justice into all environmental policies can't be understated. This encompasses the recognition of the unequal effects of climate change across different communities and ensuring equitable distribution of resources to the most affected groups.

Navigating the future of climate change requires political will, corporate responsibility, and individual action. By embracing a united front against this global crisis, we stand a chance of bequeathing a livable planet to our future generations.

Steps for Individuals Amidst the complexities of the politics of climate change, it's critical not to feel rendered powerless as an individual.

Contrarily, each person has ways in which they can make a difference and contribute to climate solutions. Strategically merging our daily actions with awareness, advocacy, and education can significantly impact the political trajectory of climate change.

First, adopting sustainable lifestyle practices can go a long way in mitigating climate change. An individual plays an integral role by reducing greenhouse gas emissions, as a large fraction of global emissions comes from power used by homes, cars, businesses, and other parts of the everyday economy (IPCC, 2014). Thus efforts to minimize car usage, being mindful of energy consumption, recycling, reducing waste, and buying from sustainable producers or products are all practical actions towards sustainability.

Our impact can further extend to the political arena through climate change advocacy and voting. By participating in climate action movements—ranging from peaceful protests to online campaigns—we can help draw greater attention to the looming global crisis, effectively mobilizing policy changes. Importantly, we can vote for politicians and policies that prioritize the climate and sustainability and hold them accountable for their actions or lack thereof. Combining individual efforts with advocacy and a political standpoint can shape policies and institutional practices regarding climate action. Lastly, fostering education and awareness about climate change, its drivers, and how it can be mitigated is pivotal to inspire actions in others and debunk any misleading narratives or misinformation. Helping people to understand the urgency of the climate crisis can prompt lifestyle and opinion transformations (Schwartz, 2019).

Strategies for Policymakers As we step into the prime battleground of climate policy formation, it might seem overwhelming for policymakers to navigate through the myriad challenges. However, strategic planning can assist in overcoming these hurdles. Policymakers must first engage in a thorough understanding of environmental science (Hovi, Sprinz, & Underdal, 2009), its various facets, implications, and innovations. Such in-depth studies and scientific expertise enable them to make informed decisions, cutting through the bedlam of misinformation and biases.

Collaborative decision making is another powerful tool. Policies should not just be products of a political faction, but should be the amalgamation of an array of perspectives — scientists, economists, lobbyists, and of course, the public at large (Olsson, Folke, & Hughes, 2008). This ensures the policies are comprehensive, alleys the fear of climate action being perceived as an elitist issue, and can facilitate a smoother implementation through consensus-building.

Lastly, policy-makers need to embrace long-term strategies, as climate change is a persistent problem, its solutions can't be instantaneous.

Policymakers need to critically assess the socio-economic implications of each policy (Sterner & Persson, 2008) and proceed with incremental changes that have enduring, positive impacts. They also need to promote transparency and accountability in the political processes concerning environmental policies to solicit public trust and cooperation. By adopting these strategies, policymakers can nudge our political structures towards delivering robust climate solutions.